Successful Business Strategies Using Telecommunications Services

For a complete listing of the *Artech House Telecommunications Library*, turn to the back of this book.

Successful Business Strategies Using Telecommunications Services

Martin F. Bartholomew

Artech House
Boston • London

Library of Congress Cataloging-in-Publication Data
Bartholomew, Martin F.
  Successful business strategies using telecommunications services / Martin
Bartholomew.
    p.   cm.
  Includes bibliographical references and index.
  ISBN 0-89006-904-2(alk. paper)
  1. Telecommunication—Marketing.   I. Title.
HE7631.B37   1997
384'.068'8—dc21   97-19473
                                                                        CIP

British Library Cataloguing in Publication Data
Bartholomew, Martin F.
  Successful business strategies using telecommunications services
  1. Telecommunication
  I. Title.
  384'.068

  ISBN 0-89006-904-2

Cover design by Joseph Sherman

© 1997 ARTECH HOUSE, INC.
685 Canton Street
Norwood, MA 02062

International Standard Book Number: 0-89006-904-2
Library of Congress Catalog Card Number: 97-19473

10 9 8 7 6 5 4 3 2 1

R0125942943
Ref   MBW

Contents

Preface

The rapid developments of telecommunications and telecommunications-based business in recent years together with the advent, the serious advent, of the Internet and similar trading networks have created an electronic marketplace with new cultural, technological, and organizational issues.

More and more businesses, government organizations, and other enterprises of all kinds are being confronted with the fact that they are now becoming information businesses first and whatever else they do second.

Telecommunications has had a fundamental role in the deregulation and globalization of markets. Businesses can now move into areas of activity and geography previously closed to them. Telecommunications is a major factor in the development of nations, corporations, and trading groups. It takes 20 seconds to carry out a simple transaction in one country or industry, but all day to do so in another.

I wrote this book because I think we treat these issues in ways that do not hang together and are too complicated for most people to understand, to use, and to enjoy.

With the high rates of change we now experience, there are benefits in establishing and sharing some underlying concepts, principles, and organization architectures that do *not* change, at least not for half a generation or so.

The book is designed to be of use to managers, professionals, and government legislators. It is directed particularly at transaction-intensive environments and at the converging industries of telecommunications, computing, information services, media, and entertainment. The book is intended to be useful for those who study, develop, and use what are usually referred to as *quality management systems*. Given the wide effects of telecommunications and work upon our lives, it is hoped that this work will also be of interest to a general audience.

If I were to articulate a *professional mission statement,* it would probably be:

To enhance the practices of work in the telecommunications-based age, so that

- My customers run their businesses or lives better;
- My stockholders enjoy (and share) good profits and capital growth;
- My colleagues and I learn from each other.

Thank you to the hundreds of people who already belong in those categories, for it is my learning from them that is the basis of this book. Thanks are also extended to those people named in the text who have allowed me to build directly upon their own words or pictures.

But particular thanks are due to my wife Tereze Svilane for her unstinting support throughout the whole process of writing this book, my illustrator Liz Hale, Colin Coulson-Thomas for writing the foreword, my reviewer, and all of the team at Artech House who have coached me. Thanks to all of my close family and colleagues who gave encouragement and advice in generous measure. Finally, the book is dedicated to two people who will live longer than I in the new telecommunications-based age: my daughters Leonie Jane and Elizabeth Rhiannon Christina.

Foreword

Information and knowledge are rapidly becoming commodities. They are all around us like the air we breath. The latest thoughts from worldwide networks of experts, on almost every subject imaginable, can be accessed and shared via the Internet or a corporate intranet.

Because it is so easily available, less value may be put on knowledge for its own sake. Those who have painfully acquired knowledge may find their social positions eroded as the emphasis is switched to how knowledge might be used to enable greater value to be delivered to customers and other stakeholders. One's information, understanding, and expertise is only relevant if it relates to the shared vision, purpose, and objectives of the enterprise. Information and knowledge need to be harnessed and utilized to contribute to both business and human objectives.

However, the competence to beneficially use and apply information and knowledge is more difficult to acquire than the ability to merely access it. We need to move on to a new era, in which the emphasis is put on relationships, relationships between people and between technologies. Shared learning and mutual understanding grow out of iterative and supportive relationships between people with shared visions and values.

Revolutions in how business is undertaken within markets, and the creation of new markets, generally require the creative combination of complementary technologies, which in turn demands an understanding of how they interrelate. Communications technologies are critical, because if the latent potential of other technologies are to be harnessed and applied people need to be able to communicate and share.

No generation in history has had so many options for earning and learning at times and places, and through relationships, of their choice, or so many alternatives in terms of how, where, and when to buy, or who to buy from. As a consequence, commercial success can be more fleeting than ever before.

Because new sets of skills are needed to play what are in essence new games, existing businesses are at risk. Why should an established business be any better than a new entrant at designing a Web page?

Loyalty can no longer be assumed, as people move from Web site to Web site like flocks of birds flitting from one water hole to another. Customers need to be retained as well as won. Sustained relationships will be those that continue to be mutually rewarding.

The backbone of Martin Bartholomew's 'warts and all' analysis of both the risks and the opportunities is a systematic and comprehensive approach to the questions that need to be asked to successfully establish telecoms-based service businesses. He graphically portrays the new 'gold rush' for electronic market space and relationship enterprises.

The emerging technologies of electronic commerce are democratizing opportunities. In order to differentiate and stand out, people and organizations need to focus on who they are and what they are about, respectively. Electronic communications will cluster around those who are absolutely outstanding. Few will bother with those who are merely very good.

Being accessible is crucial. Not to be connected in the relationship age is the equivalent of physical incarceration on a remote island. Stay relevant and keep in touch—arm yourself with Martin Bartholomew's book.

Professor Colin Coulson-Thomas,
The Willmott Dixon Professor of Corporate Transformation,
University of Luton

Introduction

We are all involved with the *Information Society* even if we do not realize it. In most people's eyes it started when computers became commonplace—perhaps starting in the mid 1970s. But the impact of computing is tiny when compared with the impact of advanced telecommunications that we are now experiencing. Not only are the telecommunications technologies advanced, but telecommunications terminals are virtually ubiquitous. Not only are telecommunications technically accessible to us, but they are fundamentally easier to use for most people than computing on its own. What is the relevance of this to business?

In many areas of business, it is the enterprises with the best approaches to telecommunications-based service that are staying in front. Will you use an airline that doesn't answer the telephone to you? Do you do business with the people who don't call back? How competitive can you be without the capability to shift around at will voice, text, and images? Without a changed view of organizational processes, structures, and cultures, many firms will fall behind.

The book draws upon examples from around the world, though its most direct perspective is from Europe, rather than America or the Asia-Pacific region. Though it *may* be easier to apply the organizational paradigm in Anglo-Saxon societies, there should be insights and approaches from which other societies will also gain. In particular, newly industrialized nations may be able to use new telecommunications technology and its accompanying organizational paradigm to leapfrog an entire phase of development.

Telecommunications, and its positioning within the economic and social framework, is a fundamental accelerator or brake for competing nations. Telecommunications-based services have been available for some time in the Nordic countries, such as Sweden and Finland. The United States was and remains the premier large marketplace. In the larger countries of Europe, France seemed to be the first to stand out with its 1980s visionary action of making screen-based services available by placing *Minitels* everywhere. The United Kingdom has the lead now with its better developed market structures and the impetus provided by its role as a financial center. Germany is beginning to show strongly. In the Asia-Pacific region Japan gains a clear advantage from its huge technological capability; Australia is one of the most innovative users of telecommunications-based services; and the mass markets of Hong Kong and Singapore have the greatest penetration of new information-based services.

The remarkable growth in the Internet is bringing people to a new idea of how business is done and how increasing portions of our lifestyle will be experienced. But the *Internet,* as first understood, has slowed down, especially at the time of day that California wakes. It has become unsuitable for much of what is now transacted across it. In the end that doesn't matter, for the term Internet is moving to be a generic one.

The Internet concept is now segmented. There is now an *Internet 2* for a select grouping of academic users who want to find, retrieve, barter, and exchange information as they have always done. Commerce is using a meta level of the Internet—the *World Wide Web* (WWW). Businesses and their partners are using WWW and other Internet technologies to build *intranets.* Business propositions are built upon hybrid technologies—such as cable TV plus games plus a public voice network return communications path. Shopping uses *electronic malls* built with Java™ applets linked with *virtual banks* and the existing socio-technical infrastructures of the credit card industry.

But, for all the focus upon the Internet, the fact is that the huge bulk of business is conducted over regular telecommunications networks. Much of it is voice-based.

Many information-based businesses deliver their services with *call centers*—handling telephone calls in a prompt, speedy, highly professional manner with a high level of automated support. Information is brought to the call center agent rather than the caller being transferred from department to department. However, service delivery can take place without call centers, in the case of low-use or highly specific services, for example. Call centers all use telecommunications, but only some call centers deliver *telecommunications* products per se. However, more and more services can be reconceived and delivered as *telecommunications-based.*

In the end, technology does not matter much to the user. If provided products and services are accessible, effective, and a good value, then the enabling computing, communications, and telecommunications technologies will be no more interesting to most people than the insides of a television set or the integrated circuits in a pay phone.

Whereas product cycles get shorter and shorter and technology changes constantly, there never seems to be as much attention paid to business process cycles and changes. There is a simultaneous need to find a new basis for organization, a model that will accommodate change while remaining stable and comprehensible. People move on; organizational trees change; emphasis shifts from one area to another (sales, say, or customer service). What does remain constant is the set of necessary processes to carry out business. This book provides an opportunity to tell people more about taking practical advantage of the new possibilities being presented.

The range of possibilities for information-based products and services is increasing, especially now those possibilities reach across several technologies and industries, such as telecommunications, computing, information services, and entertainment. This increased complexity results in a problem for customers but a business opportunity for *service providers*. A traditional role of the service provider has been as a straightforward reseller with little added value; therein lies the risk of being squeezed out of the supply chain. The opportunity is to add more value. The need for systems integration has increased and power has shifted towards systems integrators, away from the producers of basic products, such as network operators and hardware manufacturers. Service providers can create new opportunities and ensure a sustained competitive proposition by stepping up to the integration role and making products more accessible and more usable to their customers.

There is a gap in many people's practical knowledge of running businesses in this new environment. Simultaneous technology convergence and globalization of markets are bringing to the fore new ways of doing business, but our formal models of organization are slower to change. Although there is a need to change, there is also a need for the humans in organizations to see reason for change and for them to understand the overall paradigm.

Telecommunications-based service provision explains detailed strategies and methods for developing and delivering communications services of sustained competitiveness, in selected markets, built upon hardware/software products and information content from telecommunications, computing, information services, and entertainment.

The book consists of three main parts, plus a set of conclusions. The first part of the book explores the context and identifies the need for new approaches to many businesses in the telecommunications age. Chapters 4 and 5 in the second part of the book define the concept of a telecommunications-based service provider and the organizational architecture and approach that is required. The third section of the book, which includes Chapters 6–13, provides a   detailed and practical explanation of the following eight process areas, giving a tool kit for wide application:

1. Direction and strategic planning;
2. Marketing/intermarketing;
3. Creation and management of products;
4. Obtaining orders;
5. Implementing orders;
6. Providing continuing service;
7. Supporting and leading the organization;
8. Managing information.

Each of those areas is interdependent; and that is how the book is constructed. The book's scheme is to identify the components of the value chain (what happens between the inputs from suppliers and the outputs to customers); to illustrate the constituent, interdependent processes; and then to show in detail how they are engineered and why. Each process and the *set* of processes is linked to an information systems design that is developed throughout.

However, this can only be done once the context is set and the initial concepts positioned. Please move on to Chapter 1.

The Telecommunications-Based Age

1.1 WINNERS AND LOSERS

On a Saturday during 1996, between 1705 and 1730, I called BT Directory Inquiries from my home in Central London and obtained the inquiries and reservations number for 11 airlines, which I then called. The results are shown in Table 1.1. Some of them weren't there to receive my call. They wouldn't have even reached the start line if I'd been a real customer.

Table 1.1
London Survey of Telephone-based Airline Reservation Services

Airline	Number Given by BT Directory Inquiries	What Happened When I Called	Services
British Airways	0345 222111 (0345 is local rate call from anywhere in UK)	0 rings to human agent (but a just perceptible postdial delay). Open 24 hrs.	Fares, times, availability, booking and payment, insurance, car hire, hotels.
British Midland	0345 554554	1 ring to human agent. Open 24 hrs.	Fares, times, availability, booking and payment, car hire.
American	0345 789789	3 rings to human agent. Open 24 hrs.	Fares, times, availability, booking and payment.

Table 1.1 (Continued)

Airline	Number Given by BT Directory Inquiries	What Happened When I Called	Services
United	0181 9909900 (0181 is a local call from London. UA, but not BT, told me 0800 toll free call available outside London)	1 ring to human agent. Open 0700-2200. Put on hold for 10 seconds halfway through conversation.	Fares, times, availability, booking and payment.
SAS	0171 7344020. 0345 if out of London. (0171 is a local call from London)	3 rings to answer phone: Open 0800-1800/08001700 Sat & Sun. Call long distance number for urgent flight inquiries or Denmark number for urgent bookings.	
Singapore	0181 7470007 SIA, but not BT, told me of about 5 local offices outside London	2 rings to human agent. Open 0800-1800/0800-1730 Sat/0900-1730 Sun.	Times, noncomplex fares, availability, booking and payment.
Qantas	0345 747767	2 rings to answer phone. Open 0830-1900/0800-1700 Sat & Sun	
Virgin Atlantic	01293 747747 (01293 is a long distance call)	2 rings to human agent. Open 0645-2200.	Fares, timetable, availability, booking and payment. Car hire.
JAL	0171 408 1000	8 rings to answer phone. Message: " Sorry unable to deal with your call at this time" then silence until I rang off 30 seconds later.	
Air India	01753 684828 (01753 is a long distance call)	2 rings to garbled answer phone. Message: "Arrivals and departure info available form our recorded announcement...,"	

Airline	Number Given by BT Directory Inquiries	What Happened When I Called	Services
Alitalia	0171 6027111 0345 if out of London	1 ring to answer phone. Open 0830-1830/0900-1400 Sat. "If urgent reservation, hold while we transfer you to Heathrow Airport." 30 rings and then I decided to ring off.	

In the United States, the introduction of the *toll-free* call using 1-800 was a fundamental lever in changing the nature of the way business is done. Within a comparatively short time the message was getting through to customers that some suppliers wanted their business enough to pay for their calls. The United States has telephones with letters as well as numbers. There is no doubt in the American consumer's mind that 1-800-FLOWERS will provide access to a service that will deliver a bouquet to someone important to express thanks, apologies, or sympathies. The service is quick; it is efficient; it is the one remembered first. 1-800-AIRWAYS is also memorable.

Technical innovation can be harnessed with a customer service motivation to provide services that are simultaneously less costly to run and of more value to customers. Shopping by telephone is increasingly popular. When there are too many calls to be handled immediately, customers are forced to wait. It does not have to be so, for they can be connected to the stock database and order entry computer via *interactive voice response* (IVR), or indeed via an Internet server using the same interactive application. One cable TV teleshopping company displays its goods as a television program and customers connect to a call center to make their purchases. The retailer presents its IVR method as a positive method of jumping to the front of the line by using self-service. Great care was taken to coach customers in the use of the system, the scripts were developed with care, and the whole channel was presented to the call center agents as a positive way of handling routine/boring data entry tasks that left them with more time to do more interesting things.

The penalties for poor telecommunications-based customer service are loss of market share, reduced efficiency, and denial of access to new streams of revenue as new competitive propositions become viable.

The airlines in the survey that did not perform well were probably more focused upon flying airplanes than flying passengers. But even if most airlines, in many countries, are now used to competition, then rail services are largely still a monopoly. Traveling by train has many advantages, but people will often avoid it because it is so difficult to obtain information on fares, times, and availability and then to book a ticket. Generally speaking, it is quicker to go to the train station and wait in line. Cash-operated tickets vary in their efficiency. If you ride the Long Island Railroad, (LIRR) then be sure to allow plenty of time for the ticket machine. These problems will go away—or else the travelers will stay away. For a start, all of the timetables could be put on the Internet. We have the access methods.

The same patterns of winners and losers can be seen in banking, investment, insurance, and other services. Charles Schwab has a range of telephone-based services. It has a large-vocabulary voice recognition solution that lists many thousands of stocks and the key words for carrying out transactions. Elsewhere, some banks, still locked into batch processing systems, are struggling to bring the costs of extensive branch networks in line with lower cost competitors. The insurance industry has been tipped over by the newcomers using the telephone.

Governments in many countries have expended huge sums on creating computing systems to mechanize what is done in social security, taxation, and similar areas. These initiatives have been largely disappointing. Projects have extended over many years, budgets have overrun, applications have been overtaken by environmental changes. Government officials are unable to quickly provide the changes in monolithic computerized systems to meet the requests of their political masters, and their electors. Staff are as frustrated as customers when what is self-evidently required cannot be provided.

It is not just the vast transaction-intensive enterprises that are amongst the winners and losers. It is already nearly ten years since a young employee from a respected local theater booking agency left his job and set up his own business. He formed a national agency service. He built upon his contacts with theaters, concert halls, and other venues. To all appearances, he was operating in a series of local areas. In fact, he ran everything from the study of his own home with telephone links from each of the main conurbations where he'd installed a local phone number. Nowadays, of course, he could install an 0800 and let the public operators' intelligent networks do all of that for him. And that is just what he has done. Bringing together his knowledge of the business, his contacts, and his simple database of customers; he is now doing big business—without ever becoming a large organization

The Internet, even with the drawbacks it has today, is nonetheless giving access to untold amounts of information. As just one example, the Internet is a boon to market researchers; those businesses that are accessing this pool of information will beat those who are not. Those people who check out a firm's Web site before a visit are better prepared than those who do not.

Increasingly, winners in business will be those who use the Internet or similar wide area networks. That last qualification is important. The Internet, as it stands, is an orphan and without formalization it is already becoming known too for its delays and muddle as it is increasingly put into use as an almost free mass-market medium. With the exception of e-mail functionality, the Internet is still used by relatively small numbers of people to actually transact business, though that number is growing as the rising generation moves through school and college. The Internet is discussed more fully in Chapter 2.

There are increasing numbers of Internet look-alikes. In the introduction I explored the thesis that *Internet* is becoming a generic term for all wide area/global data networks.

Rethinking systems from the telephony standpoint provides an opportunity for massive improvement. Flexibility and responsiveness can be built in when the rebuilding of systems around what happens when a customer calls the enterprise—telephony access products, call center applications, integration of telephony with databases and standard applications—can all be re-engineered to meet the needs.

But it is not just the technical aspects that fall within the definition of a system. It is not even the technical system and the *human machine interface* (HMI). Nothing works if the organization and its people are organized badly or if they are not motivated well. The insurance industry has been turned on its head by the advent of telephone-based business. It is not just the computing systems that have changed. The whole equation of value has altered. Customer service is now considered in terms of telephone answering times and the ability of the person who answers to fix at least 80% of the questions that arise. It is the effectiveness of the overall socio-technical system that brings success. This book shows how.

1.2 CONVERGENCE FACTORS

The convergence and increased interaction of several factors is hastening a high-technology global arena. The telecommunications business has hardly begun to think of the implications for society and for business.

Telecommunications is the most dynamic industry in the history of the planet. Earth-shaking things are happening every day. Telecommunications is a "planetary economic locomotive" changing the face of business as we watch. An enterprise *based* in one country has its transactions handled in another. Whole industries, such as insurance, are transferred. An international bank can be run from a warehouse—anywhere.

New services and business paradigms are being created by the development of several technologies at once, technologies that were developing more or less independently but that now don't seem independent at all. Convergence in language is another vector. Businesses are spilling over into each other's areas. Economies are being protected less and opened to competition more. Political and national barriers are less of a factor than before.

1.3 CONVERGENCE OF TECHNOLOGIES

The evolution of telecommunications technology is characterized by the diagram shown in Figure 1.1. The diagram shows how telegraphy and then telephony spawned the subtechnologies of data, image, voice, audio, video, and mobile.

Every week a new subtechnology is added or there is a breakthrough in an existing technology in terms of its price, power, or accessibility. The diagram would need to be updated constantly. Note that the 1991 diagram doesn't include the Internet, *call-back* services, voice messaging, or interactive voice response. The format of the diagram was *divergent*, not convergent. It has continued to diverge. The first implication of this divergent is that there is a requirement for *increased specialization*. Similar diagrams can be produced for computing, and for film, television, games, and other technologies. Taking a selection of what is available, Figure 1.2 indicates this explosion of technologies.

This ever-growing set of subtechnologies poses an ever-increasing problem for the individual, the team, the firm, and the wider community that must spend more and more time keeping up with technology and will inevitably fail. How can these confusing and divergent pictures be managed in the mind and made useful? How can such *divergence lead to convergence?*

In fact, the scene, *as a whole*, is becoming simpler, even though each individual picture is divergent. "Anything you have seen done on *Star Trek*, except the actual transformation of atoms into bits and vice versa, is now possible or may be soon" (David Harman, Energis Telecommunications). This gives us a different way to look at technological possibility. The new paradigm has the characteristics shown in Table 1.2.

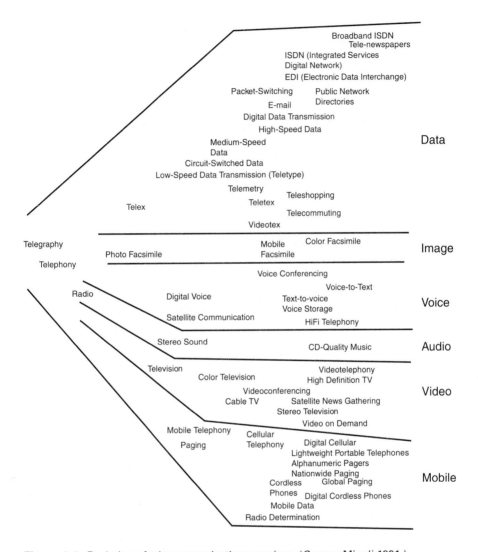

Figure 1.1 Evolution of telecommunications services. (*Source:* Minoli 1991.)

Our levels of discomfort and unfamiliarity with information technology are dropping away fast, especially as the next generation of users passes through school and college into the workplace. Although we are we still largely physical in the transport of information, we are using more and more electronics. Negroponte contrasts the transportation of bits and atoms, with an emphasis on the increasing importance of the former. Where does the balance actually lie today? How quickly is the balance between bits and atoms changing?

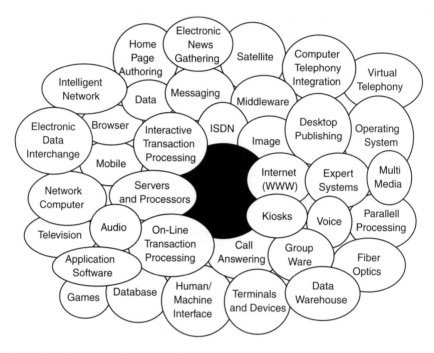

Figure 1.2 Explosion of technologies.

Table 1.2
Drivers Bringing the New Telecommunications and Computing Paradigm

Driver	*Object*
Tendency to free	Processing, transmission and storage
Tendency to unlimited	Processing power, capacity and connectivity, database capability
Tendency to intuitive	User interfaces, acceptance of product concepts, pastiche solutions

The cost of technology is said to be falling at 10% a year, and it is all becoming available to more and more people. Technology that crosses barriers between one supplier to another or from one nation to another diffuses faster. After years of work to develop internationally endorsed open standards for telephony and computing, there is genuine convergence and interworking. *Electronic commerce,* formerly *electronic data interchange* (EDI), is an increasingly common phenomenon.

E-mail is a good example of the results of convergence; it is increasingly utilized and accessible. A prospective business partner says, "Do you not have an e-mail address?" The increasingly common presumption is that it should now be the rule rather than the exception. "Why do you not have an e-mail address?" is increasingly difficult to answer with conviction. Delegates at a conference are told straight out that the presentation material and further information will not be given out as hard copy but will be available over the Internet.

Technological convergence enables, ultimately, "any client, any network, any bit and any format." Add that point of view to those of Figure 1.2 and Table 1.2 and the result is shown in Figure 1.3.

1.4 ENTERPRISE CONVERGENCE

Disparate enterprises, especially businesses, are converging too, invading each others space. Increased technological possibility is a prime reason.

The technologies used in different fields used to be different. For years, the individual worlds of business, media, telecommunications, entertainment, computing, and education developed along their own paths. Each had its own dynamics, its own vocabulary, and its own business

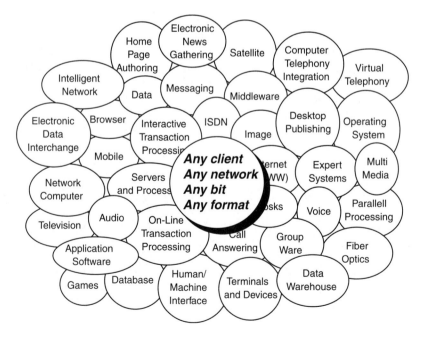

Figure 1.3 Information technologies converged.

practices; each operated in its own cultural space and in its own market. Media had newsprint and journalists, telecommunications had copper pairs and engineering, entertainment had extravagant personalities, education had resource crises and diminishing morale.

Each of these worlds had a tendency to be opaque to the people in the others. But all of that is changing now. Newspapers are produced on a computer; telecommunications companies can transport more signals faster and with fewer errors; billions of people watch a pop concert; computers are in every school, in our cars, and in our household devices. Technology knows no cultural barriers and information from each of these worlds has now flowed together.

As information technology has advanced and become easier to use, so have business possibilities become broader in scope. Entrepreneurs have seen ways to combine technologies to create new businesses. Practitioners in one business have seen ways to break into another.

New combinations of opportunities and technologies have created new niches. A new niche market is like an oxymoron at first. It is what you find at the interception of two disparate or even contradictory items or possibilities previously unconsidered as a pair. Examples are "satellite broadcasting," "telephone banking," "vegetarian sausages," and " financial information services by telephone." The development of products in the telecommunications-based convergent paradigm is the subject of Chapter 8.

Australian Rupert Murdoch, from his base in newspapers, took to space in launching Sky Satellite Broadcasting; Turner did likewise: he moved his emphasis from cable TV to the operation of the CNN Worldwide Satellite News Channel. Banks are minimizing their expensive branch networks in favor of telephone banking operations.

Alternatively, new business possibilities have not come just from a new idea, but rather the reexaminations of an old idea, or visionary desire, in the light of newly available technology. Who would have predicted ten years ago that share investors would pay a premium telephone call charge to contact a machine owned by the *Financial Times* newspaper that would provide up to the second share prices and send you annual company reports into the bargain. More and more people are buying stocks; more and more of them will never meet a stockbroker face to face. The telephone, integrated with computing, is taking over. These developments have occurred because the technology has developed to the point where they can be realized.

Telcos are rushing to link with information providers. Banks have become telcos, and retail stores have become banks. An "over-fifties" travel company reuses its database to identify and serve its clients with

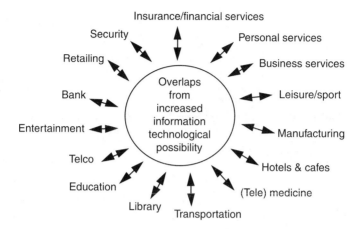

Figure 1.4 Enterprises converging.

insurance products. Airline affinity groups make special offers of wine to their members. Look at the possibilities indicated in Figure 1.4.

1.5 CONVERGENCE OF NATIONS, RICH AND POOR

Whilst it is said that 50% of the world's population has never made or received a telephone call, the other 50% already have access to a remarkably functional, accessible, fully interconnected and well-funded system. The 50% who are connected also control 90% of the world's wealth.

The provision of telephony is a major method for reducing inequality between nations and between people. Reduced inequality means reduced privilege for some and increased opportunity for others. Much of the psyche of the pioneers of the public Internet is attuned to similar visions of equality, where everyone can be within reach of all of the world's available information.

The structure *within* the public telephony industry has reflected the above notion. The telephony business has been in the hands of a small privileged group of firms, virtually all government-owned. The deregulation of the industry has given increased opportunities to other firms to run telephony businesses and decreased privilege to the original players.

1.6 CONVERGENCE OF PLACE AND TIME

Nearly free telecommunications using wideband fiber optics shrink distance. A Swiss airline processes tickets in low-wage-cost India. Excellent messaging products enable time shift that crosses zones; personal telephony helps. A person arriving on time in the office in South Africa picks up an overnight message from South Carolina and calls a colleague in South Korea on her mobile phone as she travels home by car.

Technology spreads information across geography in ways that are independent of existing national, international, or business geography. Through broadcasting, the fax machine, and the Internet we now know more of what is going on elsewhere in the world than ever before.

In practice, there is still a way to go. Not every country welcomes increased communication with others, but it is beyond the scope of this book to discuss that particular political issue. The cross-border deployment of new combinations of business practice and technology is not always plain sailing. In 1984, the introduction of a cable-television-delivered film channel to the United Kingdom was frustrated for a while by the fact that the satellite delivery of product to cable franchise head-ends could "leak" into other countries covered by the footprint. The established business regime, encompassing complex royalty arrangements already signed was that films were licensed into each country discretely—first for general release, then for video and for cable/pay TV and, finally, for terrestrial TV free showing. Nobody had considered the impact of a new technology that could not be automatically contained within an arbitrary geographical boundary.

But business quickly found a way around the problems, where there was a payback. Satellite television has spawned a new international shopping forum. If you sit watching television in your hotel room in Central Europe you can see advertisements for special offers. To complete the transaction, you are offered a return path to a local telephone number and the opportunity to use your internationally recognized credit card. Although the advertisement may be in English, the products tend to be self-explanatory, and you are guided to the relevant phone number by a graphic representation of your country's flag. This is a good example of putting together a telecommunications-based service provision operation that takes into account from the start such issues as language, different currency, and time shift.

Electronic Mail—Epitome of Convergence

A short while back I sent an Internet e-mail from the London office of a U.S. corporation to an executive in a Swedish company in Stockholm with respect to a joint proposal we were to make to a third company in the Irish Republic. A Swiss colleague had spotted the opportunity on a public database, published in Brussels, and given it to me, via my French boss, over our internal Lotus Notes system. We managed quite well, but we had to send much of the information several times over.

We are still becoming used to working across borders in Europe. Perhaps the transactions would seem simpler if we were to substitute "office in New York," "Colorado corporation in Denver" for a proposal to a "third corporation in Virginia" from information supplied by a colleague in Louisiana found on a database in Washington.

There is now an Internet service that allows you to view price and delivery for Apple computers in several places at once. The arbitrage of price, freight, and duty costs has started.

Nations vary in their capability. Partly, it is a factor of telecommunications technology; partly it is about individual skills. The average business person in Europe is still not very comfortable creating and moving electronic information. In most parts of Europe the age profile is high and the level of computer education relatively low. Substitute "office in Singapore," "Shanghai hong in Tientsin," and "Pty in Australia" from information taken by a Thai from "a database in Malaysia." Perhaps those concerned would hardly dream of matters being done any other way.

1.7 CONVERGENCE OF LANGUAGE

There is increasing convergence in the language used by the people of the world. "Nation shall speak unto Nation," says the BBC. The rise of the English language in its various forms as a first and second language is a positive vector for increased communication worldwide. English is the language of telecommunications, computing, and air travel. English is the original language of icons such as Disney, Coca Cola, and the world of popular music (including Swedish group Abba). Whether English will be overtaken as the most common global language of business by Mandarin, Hindi, or Spanish remains to be seen; it is not immediately likely. Each of these languages, plus for example Russian, Cantonese, Japanese and German, has preeminence as a common language in regional contexts. In the meantime, the English-speaking nations do have some built-in advantages over others, even if they are not necessarily the best educated in basic technology and scientific skills.

1.8 INFORMATION OVERLOAD

Convergence of technologies of businesses of time and of place has given a vastly increased ability to humans to communicate and to transact business. The libraries of the world are opened; huge amounts of information are available simultaneously.

It is not all good. We need help in sifting information, making the best decisions, and putting them into effect. Thus, there is a *brokering* or *intermediary* business for information service providers in providing an intelligent link between customers and raw information.

In time, perhaps, we'll all be on the Internet. Perhaps we'll have the skills and tools to look at data or information at different levels of detail and from different angles just by clicking on a piece of hypertext. We'll be able to select the information that comes to us and the information that is filtered out; we can read our own "Daily Me" newspaper. But for most people there is neither the time or inclination to learn how to do all of this stuff. Many people would rather invest money in getting somebody else to do the hard and boring bits.

There is a role for people who will do information acquisition, summarizing, and presenting. That is, of course, what every newspaper and magazine does. It is not possible to read every newspaper; most people manage one or two, or they get news from television in audiovisual form. There is a clear need for a summarizing service that will provide a "second-order newspaper" based upon the overall pool of data and comment available.

1.9 *THE WEEK*

One such service is *The Week*, which provides "the best of the British and foreign media—in just 32 pages." It summarizes U.K., European, and world news and repeats significant commentary. It considers a chosen subject considered in depth, includes "boring but important," "it must be true—I read it in the tabloids," and "pc watch." Sports, cultural, lifestyle, and other news/fun items are there. I have heard it described as the written form of BBC Radio 4. It certainly includes a summary of "The Archers" soap and a listing of choices made on "Desert Island Discs." Doubtless, *The Week,* like many news publications, will soon be available electronically.

Market specific "news cutting" services are moving steadily from paper-based form to electronic. Delivery to recipients can be via fax or the Internet. Less and less such material will be transported by nonelectronic means. At present, it is still normally considered that hard copy can provide a better, more attractive format, but that is changing quickly as multimedia skills and tools become more widely available.

Increasing choice in general brings many benefits but also puts a strain upon us as humans. We are supplied faster and faster with more and more information and products, so we are faced with increasing choices and possibilities. We mostly buy where it is easiest to do so, give or take on price.

1.10 THE COMBINED EFFECTS OF CONVERGENCE

In summing up, the experience of the telecommunications-based age is also the experience of convergence. Each of the convergence elements described above has an effect on every other. The result is a melting pot of differences—political, economic, and social. Increasing commonality in technology, aided by common languages, causes a reduction, accommodation or exploitation in political, economic and social differences. Telecommunications is a primary driver that is accelerating the melting process, as shown in Figure 1.5.

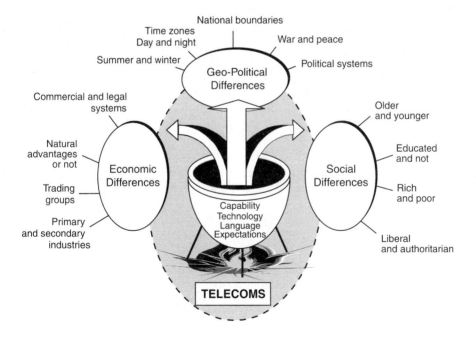

Figure 1.5 Telecommunications melt differences.

1.11 THE SERVICE PROVISION OPPORTUNITY

The telecommunications-based age and melting-pot effects of convergence have led to massively increased capability but also to complexity and a confusion of choices. This, in turn, leads to demand for a service that integrates technological and information elements.

There is a business for the intermediary that will integrate networks, equipment, and other elements into a package that services customers' direct needs. There are already many examples of this in the communications field and many of these will be addressed in this book.

Note for example: If you make a premium rate telephone call to the Financial Times Cityline, you will receive share prices automatically. The technology to do it so well was not available until relatively recently. GSM Short Message Services combined with some simple database inquiry technology will now provide yet another alternative technological solution.

People are prepared to pay extra for something that is easy to use. Every office and household has more and more equipment that is dedicated to a single task. Services are being packaged in the same way: as information and telecommunications-based direct and servicing costs diminish, their value relative to cost should increase proportionately. But it will only do so if its "time" cost to the purchaser is also managed adequately. It is not enough to provide incoming callers with a toll-free access number; they are also looking for a service that is near "time-free" to learn and to use. Successful timing of a product introduction or a working practice is not just about coming up with bright ideas before someone else does but also a function of the education level and receptiveness of people. Generally, people will not bother to learn new technological tricks if they think the learning effort exceeds the potential benefit or if they don't have time to become practiced in the mechanical steps required. They certainly won't use for long a service that keeps them hanging on or waiting in line for an answer.

How are these product and service opportunities to be addressed? Chapter 3 is taken as a convenient starting point for analyzing service provision, taking as an example the cellular airtime reseller, or *mobile telephony service provider.* Those particular service providers are sometimes thought of as an endangered species unless they add a lot more to their value proposition. The challenge for them is to simplify and retail to more tightly delineated market segments and to reduce costs at the same time.

Discussion on telecommunications-based service provision is also as relevant to straightforward customer service applications as they are to systems integration/product packaging situations. It is essential for any business to understand the impact that bad telecommunications and information handling can have on customer acquisition and customer service and to exploit the possibilities for improvement. Examples of airlines, theater booking services, banks, teleshopping, and others were given right at the start of this chapter.

The new theater booking agent was ahead of his time. The main opportunities are being enabled only now following steady advances in open standards and the introduction of effective technology for computer telephony integration. 1-800-FLOWERS is simple and highly effective. Good airlines serve people quickly and efficiently in an information-intensive transaction; they do not just fly planes and dispense food and drink.

1.12 CONCLUSIONS

Increased telecommunications brings benefits to all. Increased telecommunications provides opportunities to enterprises and individuals to transform and accelerate their businesses ahead of the competition. Increased competition is (usually and ultimately) beneficial to the general good. Competition is not the whole story; intense amounts of cooperation amongst telephony network providers is also required if the supernetwork is to function well. What of the other telephony industry—the Internet? The Internet community has been based first and foremost on cooperation, and issues of pricing have come second. Competition amongst Internet service providers is now there. Competition amongst "Internets" is also now starting. Chapter 2 looks at some of these issues.

It is generally accepted that we shall all eventually be linked together by computing and telecommunications. Carriage, processing, and storage will be nearly free. The hardware, software, and other components will be almost intuitive to use via a range of interface methods.

How far along this path are we? Communications—the movement of information from one entity to another—is changing our lives and changing the ways in which we do business. Computing—the processing and storage of information in usable formats—is doing so, too. The combined effects of increased wealth and education, technical advance, globalization, and deregulation of markets are at the heart of these changes.

Telecommunications and information management are of vastly increased significance as "factors of production" alongside the classical categories of land, capital, and labor. Communications are central to social and domestic interaction, education, and entertainment.

Telecommunications melt differences between nations, businesses, and people. The far-reaching effects of this remarkable revolution are hardly understood at all.

Consider a scenario: "Imagine a 100% tax is put upon fossil fuel—in order to make telecommunications just 1% of their present cost. Describe the effects..."

There are countless new opportunities arising for telecommunications-based service provision as a result of the continuing downward pressure on technology costs and the constant arrival of new technological possibilities that change.

We already have virtually instant communication that is location independent. We can convert voices, data, images, or money into digital form, then manipulate them and move them around at will. The parameters of time, cost, and availability are changing the norms of volume, frequency, and expected response time. We can drive things faster; those that choose to do so can gain a competitive advantage in economic terms. "Time is money."

Can *you* afford the risk of not getting your business "wired?" If your competitors are now marketing their products globally 24 hours a day, whilst you are still country-bound, time-bound, or just plain bad at answering the phone and retrieving required information, then you will already be falling behind.

Bill Gates says that knowledge workers in industrialized countries will face competition from all over the world, just as manufacturing workers in industrialized countries have faced competition from developing nations over the last decade. The combined availability of telecommunications, computing, and some common languages is forging a new dynamic working and trading culture.

The new telecommunications based age has been going for a while already. Do something now to protect market share, reduce costs, and identify new revenue streams ahead of the competition.

The Internet

2

2.1 THE INTERNET AND TELEPHONY NETWORKS IN COUNTERPOINT

More, much more of today's opportunity for mass market and business communications services is off the Internet rather than on it. But there is no doubt that the Internet has been a major factor fueling an information revolution. Arguably, it is the single most important communications development of the 1990s. Three recent developments have accelerated development. The first is the advent of browsers and search engines for easier navigation. The second is the launch of Java software to create exciting new applications whilst taking complexity from devices and into the depths of the network. The third development is what is seen as a strategic "U-turn" by Microsoft, such that this highly influential industry player is now right at the center of the network explosion.

How can it be that 100 million people will soon be using an information medium that was unknown to all but a few academics and scientists a few years ago? What is the Internet and how does it work? Why is it almost free to use? Will it go on working? Is the Internet the global information superhighway?

The evolution of the Internet has been based on three fundamental drivers:

- Its charges are independent of distance, time, and usage.
- It has an everywhere-to-everywhere, nonhierarchical technological architecture. Traffic is generally passed packet by packet.
- It started in the world of scientists, academics, and then individual enthusiasts.

But this is changing. The disciplines, structures, and cooperative fora of the telephony industry now appear less of a curb and more of an enabler to all but the most extreme and anarchic of Internet geeks. More and more businesses and individuals are joining the Internet and its off-shoots. Phone companies are moving in and stabilizing the environment with "industrial strength" offerings. The World Wide Web (WWW) and advanced interactive transaction processing techniques are just two of the features that are transforming Internet technology into a real business medium.

Far, far more of the world's business is transacted over telephone networks.

They are contrasted sharply with the Internet, having the following characteristics:

- Charges are based on distance, time, and usage (whatever the cost model dictated differently).
- The world comprises many phone company networks joined to-gether largely in tree and branch form with traffic passed from A to B over a dedicated switched path.
- Governments and big business have largely dictated developments to date—within the three-factor production model: land/capital/labor.

New telephony thinking, Internet thinking, if you will, is having a profound influence on each of these. Phone companies are increasingly competing with each other globally and in each other's markets: Prices are becoming aligned with costs. Technologies other than switched cir-cuitry (ATM, packet switching for just two) are the existing pricing and technical models of wide area telephony. On the third point, the fourth factor of "information" has been added. Telephony functionality and low-cost computing is now available to individuals and small businesses in ways that have transformed the factors of production for all time.

Traditional and Internet models of telephony are converging with each other. Near-universal telephony and computing is transforming the world. How did the Internet come to its present point?

2.2 THE START OF THE INTERNET

Unnoticed at first, the Internet resulted from a strategy to link U.S. mili-tary and academic networks during the Cold War. The Internet, Arpanet as it was known, comprised hundreds of individual computers and net-works talking to each other. Each had the same protocol and each took responsibility for receiving or rerouting any messages that arrived. This architecture produced a robust web of interconnections and alternative routes that it would be difficult to disable. One network after another was added, with links paid for out of defense budgets. The original protocols spread elsewhere. Soon there was a nationwide set of networks operating to common standards. Academic users could access each other's databases and could collect messages left for them on computers configured like mail pigeonholes—PO boxes.

2.3 TOPOLOGY OF THE INTERNET

The Internet is a worldwide network comprising national networks, local area networks (LANs), and computers. These elements and the transmission bearers are all connected with intelligent routes—computers whose function it is to recognize destinations and origins of messages and to oversee and execute the movement of those messages, error free, over the next "hop" to the next point on their overall journey. The Net is shown in Figure 2.1.

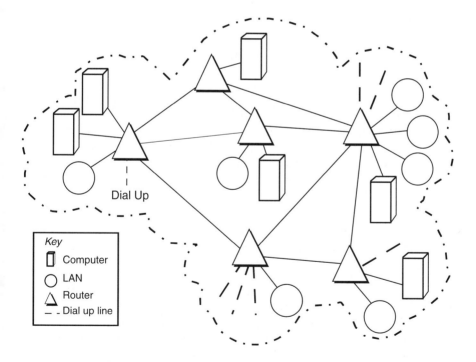

Figure 2.1 Basic topology of the Internet.

There are three principal ways of being connected to the Internet:

- Dedicated access lines—continuously connected private circuits, medium- and high-capacity (mostly corporations and larger academic institutions);
- Dial-up—for noncontinuous, lower capacity use (mostly individuals and smaller academic institutions);
- Via an Internet service provider that will seek to provide fast, reliable access by supplying free access software and a modem to connect through to their service.

Use of the Internet spread to nonacademic users, who were charged a fee. The networks spread internationally. For many years it remained hard to use the Internet unless you had well-developed information systems skills. The principal applications comprised electronic mail, discussion forums, and file transfer.

But, when the WWW came into use just a few years ago, much wider possibilities were opened.

2.4 WWW

Finding data would still be difficult without a specialist method to do so, since it is not catalogued in any comprehensive way.

The WWW is an easy-to-use feature that helps search information contained on the Internet. The Web is based upon the concept of a web site or home page for a particular body of information, such as information on, or hosted by, a particular organization. The WWW version of an address or phone number is a universal resource locator (URL). Users navigate to Web sites, access them, and then download information via a Web browser such as Mosaic™, Microsoft Explorer™, HotJava™, or Netscape™. A typical URL is http://WWW.Xcorp.com.

When you look at a home page, you see text and graphics. These documents are written in hypertext markup language (HTML). Some of these sites may be multimedia—containing not only text and graphics but also video and sound.

A portion of the text, graphics, or special icons are highlighted; this is called hypertext. If you click on hypertext, then you are linked to other pages on the Web site, or even to other Web sites on computers elsewhere; these links are called hyperlinks. In other words, the hypertext is the user interface to the URL. Pages linked to Xcorp might have addresses like http://WWW.Xcorp.com/services, http://WWW.Xcorp.com/specialoffers, or http://WWW.Xcorp.com/offices, and you would move hypertext around as shown in Figure 2.2.

It is becoming more and more difficult to find a significant organization, especially in the information businesses, that does not have a home page. Increasingly, the home page is the premier source of information to be tapped before, say, a sales call or job interview, or for general market research or product information. You can also keep up to date with stock prices or read newspapers online.

2.5 USERS OF THE INTERNET

With each month the number of users goes up and up. Many Internet users are online for more than 10 hours per week. Research on the use of the WWW suggests that sex is uppermost in the minds of many of those users. Ten of the top 25 usenet sites are dedicated to circulating pornographic pictures. Seventy-five percent of Web sites are in North America. Nearly 55% of users are below the age of 35. The largest group is the 25 to 34-year-old middle class male. Females account for just 10 to 15% of users.

But there is no doubt that the Internet is fueling an information revolution and that it is finding itself increasingly at the center of the telecommunications of a community or enterprise. The Internet is sometimes

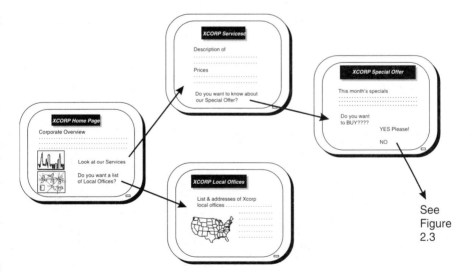

Figure 2.2 Using WWW URLs and hyperlinks.

providing educational or development opportunities for people who would not otherwise have them.

In Queensland, Australia, the Internet plays an integral part in the schooling of remote individuals and in the provision of multimedia-based information. Countries of the former Soviet Union have joined with alacrity, though telephone links are still generally poor. The University of Latvia and some of its adjunct subscribers are connected via fiber-optic cable; it will be interesting to see if this provides a discernible extra stimulus to the knowledge-based aspects of its economy.

The Internet is not all-pervasive. In Myanmar, the unauthorized possession of a networked computer is said to be punishable by 15 years imprisonment.

2.6 CONNECTING TO THE INTERNET

More and more company e-mail systems are being linked to the outside world via the Internet. Corporate and individual subscribers are often connected via commercial and Internet service providers (CSPs/ISPs) such as America OnLine, Demon, and CompuServe. Few SPs have been profitable up to now. Prices for connection and for monthly subscription will need to rise. However, the market will split into commoditized connection and value-added services. One confusion has been that CSPs (such as CompuServe) have created their own networks with services included but have not given users free access to the WWW at large. ISPs have tended to give direct connection; they give a range of support options that rival those of CSPs at lower cost.

Telcos are becoming Internet service providers, both to protect their customer base and as one part of the effort to bring stability and "industrial strength" to the Internet as a whole.

Hybrid service provision bureaus are springing up that will bridge the gap between Internet connection and the PSTN with protocol transfer/universal mail platforms—digitalmail.com is an example

Since it is becoming the norm for companies to have a WWW site, there is a growing secondary industry setting up these sites and authoring the content of Web pages.

The Internet, especially the WWW, is becoming a serious business medium, though not without difficulty. Many enterprises are unwilling to accept the drawbacks and delays of the Internet proper. Breakdowns and disasters have become much more noticeable. When America OnLine's service went down in mid 1996, over five million customers were left in the lurch until the problem was put right. Enterprises are building intranets of their own, albeit using Internet technology and linked to the outside world. A group of American universities has decided to set up Internet II for themselves.

2.7 DOING BUSINESS OVER THE INTERNET

Internet proper or other Internets, the habit of doing business is growing. The situation is still somewhat fragmented, and electronic diffusion to many more people is required.

Every few months it becomes easier, more common, and more acceptable to do carry out business transactions over the Internet as well as use it for information exchange. Something new on the Web is the "virtual bank"; try it for yourself.

There are a number of supporting activities that are making the Internet more and more of a serious business medium. From the enterprise and entrepreneurial viewpoint, there is a growing range of books telling you how to make a fortune in Cyberspace. A new word has appeared: intermarketing. For more details, see Chapter 7 (Marketing and Intermarketing).

Work has begun on such matters as intellectual property, defamation, privacy, junk mail, branding of domain names, contract formation, and money. Similarly, the law is catching up on such matters as obscenity, pedophile registers, and other matters of public concern. Approaches vary around the world, but the initiatives started by Singapore in these regards are being followed up steadily by countries elsewhere.

Already all manner of items are being bought, sold, or bartered. Note the following examples:

- Two years ago, two U.S. lawyers advertised their immigration services on Internet bulletin boards and, it is said, generated $100,000 of business. In amongst their 25,000 inquiries, they also generated a furious tirade from users who did not want to see that section of the Internet commercialized. The argument still rages.
- A house mortgage lender placed an advertisement in a newspaper. There was no postal address or telephone number given, just an e-mail address. Clearly, the lender was targeting a market of young professionals.
- In Chapter 1, the global Internet marketplace was exemplified by the service set up for checking price and delivery for Apple computers from alternative suppliers, and then ordering them.
- Every browser has its billboards—with hypertext to take you straight to a sales proposition.

Security is still regarded as a significant barrier to commerce. With every week there is a new development in firewalls and similar devices to keep marauders and viruses out. In fact, there is little difference between Internet security and telephone security. Isn't a credit card risked or compromised every time it is passed over the phone? However, 1996 was probably a watershed year when encryption became more widely available. At the time of writing, security is more advanced in the United States than elsewhere.

2.8 NEW INTERNET TOOLS

The introduction of easier to use Web browsers, such as Netscape and Internet Explorer, has accelerated the transformation taking place. So too has Java.

It is now possible to make international phone calls on the Internet at local call charges. This may be a relatively minor phenomenon at present, with few customers, but it adds to market forces that are hastening the dismantling of international cartels of the telcos that, for years, have been able to obtain high rates for international calls.

Other technological solutions are coming all the time. One such technique is the Periphonics Corporation's PeriWeb™, which is the Internet version of voice interaction with a database. Figure 2.3 shows how it works. The application on the information server, developed with the PeriProducer™ graphic user interface tool, is simply ported between the voice-based and Internet-based deployments.

Meanwhile, Larry Ellison's Oracle Corporation has announced a $500 network computer. Essentially, this will be a virtual PC—a terminal linked to the Internet where suites of applications (written in the new Java language) and multimedia capability will be found. Ellison regards the machine as future-proof and disposable; he believes it will be bundled with subscription services, just as happens with PCs and mobile phones today.

2.9 SHOPPING

Building upon the multimedia capability of the Internet, the concept of shopping malls has been developed and trailed. But it isn't the standard Internet that is likely to be used here. Various local and wide area broadband networks will allow prospective shoppers to be whisked along a

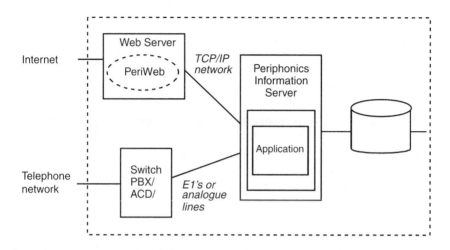

Figure 2.3 PeriWeb interface between applications and the WWW (Periphonics Corporation 1996).

Cyberspace corridor or street until they arrive at a shop (Web site) they wish to visit. The process is then repeated inside the shop until the shopper finds an item of interest. Hypertext links provide access to detailed information or video demonstrations.

Electronic payment is made using credit card or electronic purse and the purchased item is delivered to the required physical location.

The shopping mall can be brought into the home. Section 1.1 described a narrowcast shopping channel on television (broadcast shopping programs, for that matter) bring goods visually to the potential customer. They are presented, described, and extolled over the television and the purchase is made over the phone. Conceptually similar to the mail order catalogue, this simple development adds real-time excitement and a personalized appeal to the shopper by a skilled sales presentation. The shop can buy in a special line and then sell it all out in a single night. Inevitably, the order lines can become clogged, but the use of interactive transaction processing means that shoppers can use "self-service to avoid the waiting lines." But another way to make the purchase is to call up a Web site and enter the order directly into the stock and dispatch computerized system, using a tool such as PeriWeb, described earlier. Security concerns, as outlined earlier, can, in any case, be overcome by pre-enrolling customers so that they do not have to give credit details "openly" over the Internet.

2.10 THE OPPORTUNITY FOR SERVICE PROVISION IN RESPECT OF THE INTERNET

The Internet is becoming more complicated, not less. It will be some years, even a generation, before the concept of direct keyboard interaction is readily accepted by most shoppers. In any case, why should people have to use a keyboard when voice-based transaction techniques and voice-text-voice conversion protocols are advancing so fast?

For those who want to learn about the Internet in general, there are growing numbers of Cybercafes where you can try it hands-on. University/college students connect to the Internet free. Internet connections will be available from all schools before too long. But there will probably always be a role for a helpful intermediary who will "take you to the shops" and "serve" you. This is certainly going to include network "angels" who provide "help" in the normal way and in better and better, more intuitively based ways. Intermediaries will draw your attention to what is being offered and Intermediaries will identify the most likely customers for what is being offered.

Intermediaries will guide and explain and make the bridge between the customer and the elements of technology and information that are required to complete the chain to the product. This is the new retailer. Sometimes it will be automated—the deployment of a preformed expert system response or helpful prompt. Sometimes it will be human.

In the same way, there will be a continued and intensifying need for intermediate consultants, suppliers, integrators, and maintainers of products and information services for businesses.

Whether or not much of this function is carried out by automated means, initially it will be a human who acts in these helper roles. This has been characterized in earlier times as a retailer. In the information age, the intermediary is a telecommunications-based service provider, including Internet service providers. This first generation of Internet service providers has, however, done little more than provide a service of connection and access to a range of information services.

A paradigm shift in Internet service provision is coming. Some providers are moving closer to the customer and further away from the technology. They are concentrating more on providing a business advantage or lifestyle necessity or enhancement and less upon the technology that is being employed to do so. Though the technological descriptor *Internet service provider* will eventually become irrelevant, there will be a need to provide help in that domain for a long time to come.

The perspective of the service provider is that the Internet and its derivatives are amongst the transportation and storage devices available for use. The Internet itself is not the only component in a telecommunications-based service proposition.

2.11 CONCLUSIONS

Even if the growth of the Internet is one of the most talked about manifestations of the changing communications paradigm, it is certainly not the only one. Despite its phenomenal growth, It is estimated by some that there may still only be between 25 and 30 million real WWW users at present. At the start of 1997, it was estimated that there were 70 million e-mail users. The proportion of users for business purposes is growing rapidly, but it is still a relatively small number. Of more significance, the number of users using technologies that include the Internet is larger by a factor of perhaps five. (e.g., e-mail).

We are still only at the start of doing business on the Internet, though matters are advancing rapidly, especially as more intermediaries come into being providing easy-to-use services at one remove and especially as shopping and other business becomes a reality. It has been estimated that there will be over 200 million people connected to the Internet by 2002. But it will not be the Internet as we know it today; it will be a constellation of networks like the Internet or like the telephony networks of today.

From the viewpoint of such Interneting as a business medium, the key accelerators will be acceptable forms of money transaction, increased use of intuitive interfaces, and continued rebalancing of transaction cost models *towards* bits and *away from* atoms.

On or off the Internet, there is a huge opportunity space for the exploitation of telecommunications-based service provision. Customers can reach suppliers, and their competitors, too, more easily than ever before. The lesson of the Internet is that low-cost accessible telecommunications, data manipulation, and storage create a revolution that few established business socio-technical systems can withstand.

But when all of the serious analysis is finished, it still has to be recognized that the Internet and the WWW are now being carried along in a surge of excitement that shows no signs of slowing.

Providing Mobile Telephony—Case Study

3

3.1 CASE STUDY INTRODUCTION

This is the third and final chapter on the *context* of telecommunications-based service provision. Building upon the examples in Chapters 1 and 2, the case study of U.K. cellular telephony service provision provides a vehicle to describe and analyze a real situation so that inferences can be drawn to assist in developing a general conceptual model.

The United Kingdom was chosen for study in this chapter because it is unusual in having *legislated* for service provision to be an integral part of a new market and technology right from the start.

Although mobile communications have been available for some time, cellular mobile telephony is relatively new—little above 15 years old. Table 3.1 introduces the main components of the cellular telephony industry needed to operate a mobile network and provide services upon it.

Table 3.1
Components of a Cellular Telephony Industry

Function/Process Area	Methods and Examples
Build and operate network. Provide dial tone and operator services.	One or more licenses granted by government PTO extends existing network infrastructure New market entrants
Contract and bill for airtime * e.g., calls and the contractual arrangements to pay for them.	Network operator Service provider Third party: selling to its existing customers/ members Dealers and retailers on behalf of the above

Table 3.1 (Continued)

Function/Process Area	Methods and Examples
Buy and resell handsets.	Network operator Service provider Dealer/distributor
Provide value added services e.g.; voice messaging; short message services; private circuits; accessories; car kits and installation. Short code dial for a taxi service.	Network operator Service provider Other third parties

There are three main methods of modeling those components in actual situations:

- Networks built and operated and handsets and customer service added to existing technical and commercial infrastructure of public telephone operators (e.g., United States, Finland, Hong Kong).
- Networks built and operated and handsets and customer service provided by new market entrants with a range of shareholders (e.g., Vodacom in South Africa, Optus in Australia, Belcel in Belarus. Additionally, each of the countries in the list above has now extended its licenses to two three or more licenses in each area.).
- Networks provided by PTO or new market entrant that are prohibited from providing services direct to customers. Handsets and customer services provided by *service providers* who buy airtime and resell it to customers (e.g., the United Kingdom, initially).

3.2 SERVICE PROVIDERS AS A PROTECTED SPECIES

Clearly, it is the third model that is of particular interest. Mobile service provision was ordained by the U.K. government for the mobile industry as an act of industrial/competition policy. This structural convenience, ended in the mid 1990s, clearly did no harm to the early fortunes of service providers.

But the squeeze on margins in the telecommunications industry finally hit the service providers hard, and the demise of many service providers is regarded as inevitable. Ironically, however, the advances in technology of the last 10 years have created many new opportunities to create sustainable, competitive propositions better than those that have gone before.

In 1984, when the original cellular licenses were granted in the United Kingdom, the two new network operators, Vodafone and Cellnet, were not allowed to deal with their customers directly. So, *service providers* bought airtime from the operators at a discount, and it has been they who have provided handsets, billing, and other services.

It is generally held that service providers did much to expand the mobile telephony market during the 10 years that they were a protected species, but that they have not done as much as they should have done to enable their evolution from simple middlemen into more competitive enterprises once that period came to an end and the commercial and technical dynamics of the market changed.

3.3 EARLY U.K. CELLULAR MARKET STRUCTURE AND PRACTICE

Table 3.2 builds upon the theoretical Table 3.1.

To customers, the entire picture was confusing. Just whose customer were they supposed to be? Customer dissatisfaction levels were high.

One of the reasons for this was a quirk of the mobile cellular market place: Its four main parties had different, and often conflicting, objectives and paths to success.

The primary interests of the *network operators* were growth and market leadership. As duopolists, they were able, anyway, to avoid competing on price and thus keep margins high enough to ensure handsome returns on their initial investment. With the financial muscle that this gave them, they provided connection bonuses to service providers of over £150 ($225 to $250) per new customer. Network operators also plowed large sums into advertising, promotion, and training.

Service providers were carried along by this structure. They paid over to their dealers a new connection commission and supplied handsets below cost. Competitive pressures indicated that service providers would pay over at the outset more than they received from the network operator as a connection bonus, for the dealer would, simultaneously, deliver back a customer contract for a minimum period calculated to provide a respectable payback.

Dealers, thus incentivised by service providers, were motivated to churn customers as quickly as possible—that is to sell them the new, latest, handset and collect a new connection bonus. Dealers would also use the old handset to gain a further customer and collect a further connection bonus from whichever service provider was currently paying the most.

Table 3.2
UK Mobile Phone Market Structure 1984–1993

Network Operators *(Vodafone, Cellnet)*	*Handset Manufacturers* *(Motorola, Nokia, Ericsson, Panasonic, NEC)*
Built and operated the networks Connected calls - including to and from other networks	Manufactured handsets Supplied service providers and dealers
Paid substantial bonuses to service providers for new customers and for net growth of subscriber numbers	Promoted their individual brand of handsets, size, battery life and style as the product being bought by customers
Sold airtime to service providers at a discount - by providing call records for each customer	
Provided messaging services private circuits and later some data products.	
Provided incentives and training to service providers	
Promoted the convenience and coverage of their networks as the competitive proposition being bought by customers	

Service Providers	*Dealers*
*(Talkland, Martin Dawes, BT Mobile, Astec, Cellcom, Carphone—later Mercury Communications Mobile Services, People's Phone, Vodac etc., etc. Also: Cell<u>hire</u> etc. Also F*ord)	*(Carphone Warehouse, London Car Telephone, many many small businesses in shops and in small workshops.)*
Found customers and signed them up for a period. (directly or via a dealer)	Bought in handsets from service providers, specialist dealers, and (some dealers only) on the secondhand market
Bought handsets in bulk. Sold handset and accessories(directly or via a dealer)	Signed up customers for service providers for which they received substantial bonuses to subsidize...
Bought airtime and value added services, as above, and sold it to end customers	...handset sales at heavily discounted prices
Provided customer service	Provided some customer service
Promoted all three of network, handset, and the SPs particular tariff or service as the competitive proposition being bought by customers	Promoted heavily discounted handsets but often would minimize the need to sign a service provider contract to get the bonus

Manufacturers of handsets were motivated to shift as many new pieces as possible, at the best possible price to be obtained in each part of the (for them) global market.

3.4 DYSFUNCTIONS OF THE EARLY U.K. CELLULAR TELEPHONY MARKET MODEL

The market model was wide open to further unintended, unforeseen, and often dysfunctional results through a cocktail of opportunism, legitimate protection of self-interest, sharp practice, and outright criminality:

- Service providers would buy handsets at the U.K. market price and resell them into other markets where prices were higher (manufacturers would seek to track this activity and refuse to deal with offending parties).
- Entrepreneurs would buy up 10, or so, mobile phones at the rates subsidized by service providers, and resell them abroad. No contract would ensue, for a false address would have been given.
- Network operators naturally ensured that their *gross* connection bonus scheme was linked to a further scheme that measured *net* growth of subscribers after disconnections. Although this had a broadly beneficial effect—encouraging service providers to reduce churn—it also discouraged the write-off of spurious contracts, so subscriber numbers tended to be overinflated.
- Handsets would be stolen and then recycled into the system by unscrupulous dealers (mostly, eventually prevented by the use of *electronic serial numbers.* At one point, 12,000 phones a month were being stolen.)
- Stolen handsets were *cloned,* so as to be electronically identical to handsets in use. Not only did this mean that the thief obtained free phone basic calls, but thieves could also use them for mass calls to premium rate lines, whose operators would be paid a share of the rate by the supplying telco. (Where fraud was proved, the operators were punctilious in writing off charges.)
- The mobile phone, with its inherent characteristics proved an invaluable accessory for drugs dealers and prostitutes—often with mobile phone theft being a source of cash to buy the wares of the former.
- Eavesdropping on analog mobile phone conversations became a new invader of privacy—with several notable incidents involving the rich, grand, and famous.

Many of the loopholes were filled one way or another, but a somewhat "seedy" atmosphere was never far away from much of the industry. Fraud, in particular, is nothing new in telecommunications, as in any transaction-intensive industry, but the mobile phone industry struck some particular chords in the public mind. Meanwhile, the "yuppie" era was raging, and mobile phones became one of the most frequently portrayed, negative, and overpriced icons of a brash and selfish age.

The United Kingdom's Chancellor of the Exchequer, Norman Lamont, branded the mobile phone as "one of the scourges of modern life" and promptly put a tax on phones provided free for private use. This tax, and both the negative and positive "status" issues, tended to slow down the spread of the mobile phone as a tool rather than a toy.

What was the impact of all this on service provision? Much of the financial effect of these dysfunctions bore directly upon the often maligned service provider. Having taken the connection incentive and paid it on in the form of commissions or handset subsidy, it was commercially essential for the service provider to extract the maximum possible margin—through plenty of calls, low bad debt, and a long contract period.

3.5 RESPONSE OF SERVICE PROVIDERS

Classical marketing insights would dictate providing a service that met the needs of customers and made them happy to stay with their supplier. Some would take a "low-cost, no frills" route; others would opt for a "premium price, special (and specialist) services" route. In a totally elastic market with "frictionless" movement of customers between suppliers, that whole scheme would have worked really well. Market forces will ensure that it does so in the longer run.

But the short-term market forces were at odds with the longer term market forces. On the whole, service was rather poor and provided few innovative product features. One commentator said that "one of the U.K.'s most successful service providers still measures customer care not in the percentage of calls answered in three rings, but in the percentage of calls answered at all" (Arthur D. Little—"Mobile Service After Amex"). In the same article, most of the tariffs on offer were variations on the same theme. Life seemed to be built around the latest handset rather than a proactive and evolutionary engagement in the real use to which customers were putting the service they were buying and the value they were obtaining. Portrayal of the mobile phone as an essential business tool tended to be general and superficial.

Besides, many of the service providers went a further stage in protecting their commercial interests. Some chose to maximize profit by engaging in commercial bullying of their customers—locking them in with complex charging structures. Ironically, some of these were wrongly calculated and actually unprofitable for the service provider. It is said that 50% of the tariffs of one well-known service provider made it a loss. High-use customers subsidized low-use customers for years. Contracts were riddled with small print and with punitive cancellation arrangements. Service providers prevented subscribers from taking their numbers with them if they left. They even refused to release their electronic serial numbers (ESNs) to prevent leaving customers from getting service elsewhere for the phones they legitimately owned. "Per minute charging" was commonplace, whereby, say, you were charged for a two-minute call the second you went beyond one minute.

The service provider faced a dilemma. To be competitive, the utmost financial benefit had to be obtained from each customer and each entrepreneurial opportunity. To provide good service, more investment had to be plowed in. But, generally speaking, the service providers were small- to medium-sized businesses without the cash available to invest for the longer term.

3.6 REACTIONS OF CUSTOMERS

Even facing some extremes of poor service and commercial cynicism, customers found it was not worth rebelling. Many users, probably most, at that stage of the market, had their phones paid for by someone else, or at least had them as a business expense deductible for tax purposes.

Corporate customers did rebel. Their specialist telecommunications managers, well-briefed by the technical press, learned how to use their buying power. If a service provider even wanted to *receive* a request for proposal, it might have to undertake in advance to comply with arrangements that removed these deleterious conditions. The popular press took up the cudgels of the consumer and warned against small print.

Many people would simply opt for BT, Mercury, or another name that they knew. But branding was a problem. What was the outward brand—Vodafone? Vodac? Cellnet? BT? BT Mobile? Mercury? Carphone—later Mercury Mobile (and later a new enterprise Mercury One-2-One, too?)? London Car Telephones? Carphone Warehouse? People's Phone? NEC? Ericsson? Motorola? Whichever *brand* you chose for safety, the customer would inevitably receive confusing and conflicting messages and information.

Very few people indeed understood the structure of the industry. Confused thus, their level of trust was low and their fears were fed by the overhangs of the negative factors already mentioned.

3.7 MARKET CHANGES (1994–1997)

3.7.1 Broadening of the Base

However, the market moved on. In 1994, there was an advertisement showing a country midwife using a mobile phone—not just to find her way to a remote farmhouse in the middle of the night, but also to remain in contact with her husband. The main motoring organizations, too, promoted mobile phones as a security device, especially for women alone in their cars. Here, at last, was an organization that understood your needs, and with whom you already had a relationship. Plumbers and others everywhere became connected. Mobile phones moved from being a yuppie icon and started to be regarded as an essential accessory for an increasing range of young urban persons. The networks introduced low-user tariffs.

All of this was beneficial to the service provider. But without a fundamental transformation of the way they did business, they would find themselves going steadily to the wall on account of five inescapable factors:

- Their generally poor service in the eyes of their customers (already covered);
- Continuing downward pressure on tariffs;
- The end of the prohibition on network operators selling directly to customers;
- The entry of retail chains into the market;
- The arrival of digital mobile telephony and new technical possibilities.

3.7.2 Tariffs

For the first years of the U.K. mobile telephony market, all parties enjoyed high margins—said to be over 50% for the network operators at one point.

Downward pressure came from two fronts: falling prices and new competition. General pricing levels in telephony had been falling for some time; mobile call charges began to look too unattractive by comparison and user resistance resulted. New operators coming into the market—Mercury One-2-One and Orange—set a new lower level of prices. The former even offered free calls in the evening. (This proved exceptionally popular though not always profitable.)

As well as competition on coverage and network quality, there was, at last, competition on prices, terms, and conditions, and there was at last a very wide range of channels to market. With a vengeance, the networks dipped into their margins to project their branding. With the drop in prices came a drop in the gross amount of the margins available to fund the functions of service provision. Most found themselves down at break-even or making a loss.

3.7.3 Direct Sales to Customers

To make matters worse, the service provider might not get a chance to enjoy the increased numbers of customers buying and using phones. Both the new networks were intent on serving customers directly for themselves.

One-2-One went out of its way to press home the proposition of dealing direct. Unfortunately, some people said that its systems and procedures did not match its marketing. But while One-2-One was experiencing "a staggering 41% churn.... Cellnet was at 32%, Vodafone at 27% and Orange at 18% and improving" (*UK Sunday Telegraph* 17 November 1996).

Orange, it is, that has changed the rules. Hans Snook, Group Managing Director , says Orange has kept its handset costs higher in order to make sure the customer looked behind the small print before signing up. Orange offers innovative tariffs, a highly functional voice messaging service for every customer, and a simple proposition of customer service from one place. Service providers are a welcome adjunct to its sales force, but they only sell handsets and a contract; they have nothing to do with pricing, billing, and backup services. Service providers have become dealers.

In the same article, John Karidis of Analysts Kleinwort Benson says, "The service providers produce more than 100 tariffs and have an affinity for the small print."

Faced with these pressures, many service providers are finding it hard to pay their bills. With the greatest of these being the airtime bill from its network operators, it is not surprising that the operators are gradually taking the money to swap it into equity in the service providers.

3.7.4 Retail Chains Arrive in the Market

After the first few, heady years for service providers, entrepreneurs identified the financial dynamics of the dealer to be better than those of the service provider. Far better, they decided, to take the connection bonus and to leave the issues of shrinking margins to the service providers.

The phenomenon of the super dealer prospered for some time—specialist retailers in the market ahead of the high street chains. As they grew in strength, they began to create the longer term face-to-face relationships with customers that the service providers had allowed to elude them. Eventually, too, they began to iron out the negative issues created by long-term contracts and to insulate customers from indifferent service. It became possible to walk into Charles Dunstone's Carphone Warehouse and buy a phone and connect it with no minimum period. With a "stock" of live service provision contracts in the name of Carphone Warehouse, portfolio effect worked to mitigate possible losses. In many ways, the super dealers became the new providers of real service.

Retail chains, such as Dixons, began to enter the market for themselves. The mass market had been primed with low-user tariffs and with the highly subsidized prices of handsets. With retail chain volumes anticipated by the manufacturers, the prices came down again.

Despite subscriber growth around 25 to 30%, the overall picture has become less and less favorable for service providers. New customers are spending much less. Long gone are the average revenue per subscriber figures of £500 ($750 to $800) per year.

Inevitably, the retailers looked at the (squeezed) performance of service providers and found them wanting, but they still needed them in the piece for they only wanted to shift equipment and not enter service relationships. No longer enjoying high prices and high-volume users, the service providers found themselves desperately fighting for market share—thus putting further pressure on already plunging margins.

3.7.5   Digital Mobile Telephony and New Technical Possibilities (1993 to 1996)

Evolution of technology is inevitable. Phones have become smaller, cheaper, and easier to use. There is an increasing number of mobile data applications. Fixed/mobile integration is more common, especially by use of a private circuit from the mobile network to the corporate private network. Network software has cut down fraud. Customers are starting to use more digital phones. The next generation of customer care computer systems, call center technology, computer telephony integration, and database marketing software are all now available, though many of today's service providers will not be able to afford them in time.

Maybe the advent of digital mobile telephony, as such, had an effect on the position of the service provider; maybe it did not have a direct effect at all. Time will probably show that it did.

Digital technology certainly gave a fillip to the hitherto slow development of mobile data business solutions. The Vodafone Group, wisely, had formed Vodata as a type of service provider to address this market. Mercury Mobile Communications Services, too, had taken steps in this direction—especially as its product management converged with its parent fixed network company. Elsewhere, it was left to individuals to cobble together ad hoc solutions on demand. Two or three analog networks and some private mobile-radio-based solutions found themselves overtaken by the possibilities of digital mobile telephony based upon GSM. Especially relevant is *short message service* (SMS).

Suddenly short message services came of age. In November 1996, the launch of a new stock price service marked a sea change for mobile telephony: The customer specifies a list of stocks and price breaks in which he is interested. The stocks are tracked and the information passed over the digital network. Unlike the pager-based or premium rate telephone inquiry services that preceded the service, both an alert and a return path are inherent.

Who needed one of the existing mobile service providers to set up such a service? What value could they add in technology, marketing, or customer service? They had missed the boat here long before. And yet, a boat they need not have missed.

Tele-Finland set up a service for mariners whereby a customer calls an interactive voice response to request reports from nominated coastal weather stations—delivered by SMS. Tele-Finland Tele-Media is a service provider that starts from the standpoint of a customer need and uses several technologies to deliver a solution.

The very size of mobile service providers dictates that they will need partners to extend their value proposition:

Back in the UK, Unisys put together a solution with BT Mobile and with Cellnet to support a mobile work force. The components of the solution were a GSM phone, a laptop computer, some middleware, and some work flow applications.

3.8 CONCLUSIONS

All of these developments provided the traditional service providers with deepening problems. Similar issues confront smaller telcos in the face of global competition invading their former monopoly arena.

The moment of crisis had arrived. The implications for service provision can be illustrated by an analysis of the strengths, weaknesses, opportunities, and threats for the U.K. mobile service provider at that point, as shown in Table 3.3.

For those service providers that were not unduly hampered by the types of weaknesses shown above, the opportunities to move ahead in a new form of telecommunications-based service provision were considerable.

Table 3.3
UK Mobile SPs: Strengths, Weaknesses, Threats, Opportunities

Strengths	*Opportunities*
Customer base (and database)	Continued growth of the market
Going concern: premises, trained and experienced staff, systems and processes	Sell out to network operator
Generally entrepreneurial attitude	Work with partner who can extend value proposition (N.B. with a fixed network operator)
Brand (possibly)	
Familiarity with mobile communications	Re-brand in association with a partner (e.g., Blue Chip X Service Provider)
Existing partnership relationships	Find new products for same market (e.g. data, value added services)
	Find new market for same product (overseas)
	Find characteristics of best 20% of customers; create heavily enhanced value proposition for them at a premium price. Dispose of other customers and scale down operation accordingly
	Do new analysis of strengths and redeploy into another form—e.g., a super dealer
Weaknesses	*Threats*
Lack of cash	Decreasing margins
Unprofitable	Hostile take-over
Weak negotiating position	Increased competition from other suppliers of core product
Lack of system/solution skills. Underinvestment in systems	Inaction during time remaining
Brand (possibly)	
Insufficient scale to compete with network operators	
End of protected position	

It is clear that any competitive proposition needs to be placed within a sound concept and architecture and then executed with vigor. Having set the context, it is to the *concept* of this new form that the second part of the book is dedicated.

Opportunity, Scope, and Definition

4

4.1 CONCEPT OF TELECOMMUNICATIONS-BASED SERVICE PROVISION

The purpose of Chapter 4 is to analyze the concept of telecommunications-based service provision—opportunity, scope and definition.

New and converging technologies, particularly in telecommunications and computing, are spawning new capability. Increasing wealth, education, globalization, and the deregulation of markets are creating virtually unlimited demand for information-based services.

Telecommunications-based service provision is the method by which many of these services will be linked with customers.

Telecommunications-based service provision represents three opportunities:

- A channel to the right markets for providers of technological and information elements;
- A method by which individuals and businesses can access technology and information more easily and get more value from them;
- A valuable business for the intermediary.

4.2 EXPERIENCE SO FAR

Fifty percent of the world's population has used a phone. Very few people, still, have used the Internet. However, the Internet has had several breakthroughs over the past five years—the World Wide Web (WWW) (especially via browsers), electronic mail, and now the availability of an object-oriented language, Java, that is stimulating the production of business applications. Internet use is, admittedly, growing fast, but its user population has by contrast been dominated until very recently by people from relatively narrow segments of the overall population: the computer literate, the enthusiast, the younger male, and the academic. Without help, without simplification, and without a better thought out value proposition, it would have been hard for the Internet and similar telecommunications media to penetrate beyond those segments. Internet-based products are now coming of age—now that the channel to them is opening up.

For all its problems and despite the dysfunctions of the United Kingdom's mobile telephony market model, there is little doubt that service provision has played a major role in stimulating mobile telephony and helping it to achieve the penetration it has done today. The mobile industry has succeeded in bringing its products into both the business and the consumer markets, and it has grown these markets to commercial scale.

There is much to be learned from the entrepreneurs and managers who were a part of that period. The story from the mobile telephony industry illustrates some principles of the service provider concept. Mobile telephony service provision has recently been working through a crisis, but is facing a new opportunity. Internet service providers struck out into unknown territory, and with care and focus, they will survive and grow stronger.

4.3 WIDER OPPORTUNITIES AND THREATS

There is more to telecommunications-based service provision than is apparent from these examples alone. There has been a major change in the way that many business functions have to be conducted.

Three major propositions of Chapter 1 were that:

- The telecommunications-based age is here now.
- There is a key strategic need for businesses in general to use telecommunications and computing effectively in delivering services to their chosen customers; businesses must act now to avoid falling behind.
- There is a role for intermediaries, telecommunications-based service providers, to help us access and use information.

The development and convergence of information and telecommunications technologies are breaking down barriers to market entry. Data processing, storage, and carriage are all tending to the nearly free and information is becoming an increasingly significant factor of production. There is an increasing predominance of knowledge and information in the value chain. Telecommunications is melting differences (see

Section 1.10). Everyone is getting into telecommunications and into everyone else's business. The examples that follow illustrate the following factors:

- Reconceptualization of enterprises as information businesses—leading to improved control of transactions and knowledge of past transactions, relative profitability, trends, and outcomes;
- Increasing importance of telecommunications as a channel to market;
- Leveraging/reusing customer databases.

4.4 RECONCEPTUALIZATION OF ENTERPRISES AS INFORMATION BUSINESSES

Enterprises are now recentered around information management. A classic network business is logistics. Management scientists think about linear programs to determine the optimum number and location of depots to serve a given market. But, whilst the trucks and depots are vital components and it is important to obtain the underlying physical cost advantages by solving the above equation, the real differentiator between one logistics business and another is the relative effectiveness and agility of its information systems. Take, in particular, the specialized logistics business of document courier services. Even if the function of a logistics business is to move atoms from place to place, the core skills are now in the moving of bits.

It can be argued that even an air defense system has now evolved into a large telecommunications network with powerful information processing capability...plus a small number of interceptor aircraft. You can no longer rely on sheer weight of numbers of aircraft to carry out the air defense task; the aircraft and crews are much more sophisticated; they are much more expensive. They need to be used with more intelligence, but you can get by with fewer.

The core proposition of airlines has been similarly shifted. The basic assumption about an airline is that it is engaged in the physical carriage of people and freight from one airport to another—the emphasis being on the flying of airplanes. But most airlines operate much the same aircraft to established technical and safety standards; their attentive staff undergo similar training; the food and in-flight entertainment are broadly similar. It can be argued that the real differentiator in modern airlines is information management—leading to the efficient linkage of passengers, fares, seats, load capacity, takeoff slots, baggage, telephone answering, weather, spare parts, and maintenance schedules. The airplanes and the flying are almost incidental. What is important is to get to your destination on time without hassle. What often feels nicest for a customer is to be treated as a known customer with known preferences.

There is much in communications service provision that is familiar to leading airlines. Air travel was once run as a white-knuckled adventure in a flimsy machine, albeit mitigated by attentive cabin staff. Things have changed; successful airlines are now run as information-based service operations. It is the successful execution and linkage of the processes of the whole journey, from booking to arrival (with luggage, too), that is at the heart of an airline's success. The flying is taken for granted. It is the superior management of all the information and knowledge that is now the differentiator and the core competence.

4.5 INCREASING IMPORTANCE OF TELECOMMUNICATIONS AS A CHANNEL TO MARKET

It was a selection of airlines that featured in Chapter 1's survey of telephone answering performance. The level of effectiveness of this form of providing information to customers is a key determinant of sales and customer satisfaction. As an example, consider the case below.

A leading car rental company expends great effort in recruiting customers for its loyalty scheme for favored customers but then fails to capitalize on the effort it has expended. Telephone this organization to rent a van. It is difficult to find the number, to get through to the right location, to obtain a consistent view on what it will charge, to excite in its employees any real desire to fulfill the needs of this (presumably) useful customer. You could even be, simultaneously, the registered driver of one of its longer term contract hire vehicles. It does not trigger a reaction, for this company is wasting its information assets! Its staff advise you that the premium customer scheme is only designed to enable a prompt drive away when arriving at selected rental locations. The company's vehicles are new and clean but its information management lets it down.

There can be few information management situations more challenging than the provision of ticket and timetable information to rail travelers. In the United Kingdom, even at the moment, there are reported to be 20 million incoming calls per year. Until very recently, the existing arrangements were inadequate, with low customer satisfaction, stressed staff, and a low confidence in the information provided. An example of the difficulties experienced was the recent publication of national timetables and the subsequent multiple amendments. It is still extremely difficult to complete the overall process of finding the right number, obtaining an answer to the telephone, and structuring one's inquiry in a form that elicits a valid and complete set of alternatives for decision about such factors as time, availability, and price. It is quicker to go to the relevant station in person.

As this book goes to press, however, it has to be said that the situation is improving greatly.

Such shortcomings also create employee dissatisfaction, which exacerbates matters further for the customer (see Section 11.1).

A sea change has come to the retail banking industry as a result of the establishment of telephone banking operations. The insurance and financial service industries have seen equivalent change. Whole networks of branch offices and the employees therein have been swept away in the face of telecommunications-based alternatives.

Recent advances in the Internet's technology have been accompanied by an upsurge of entrepreneurial flair and opportunism. Once having put information on to a home page, it is a short step to start selling products.

4.6 LEVERAGING/REUSING CUSTOMER DATABASES

The following two examples—concerning Saga and the U.K. Automobile Association (AA)—demonstrate how the reuse of information in customer databases and the development of telecommunications-based channel markets have created opportunities in previously unconsidered business areas and even transformed the very business itself. Clearly, these new market entry opportunities constitute an equivalent threat to incumbents who fail to shift to the new telecommunications-based paradigm.

Saga has been known for 20 years as the holiday company for old age pensioners. But over the past year or two it has been promoting itself as a provider of insurance to the "50s and over" market. This would appear to come from an astute reassessment of the value of its database of established prosperous people and a timely capture of the post-World War II "bulge" or "baby-boomer" population. Saga advertises heavily on Classic FM radio™, and it offers access to its products via a telephone number. Saga is the intermediary for these services, not the primary producer of their basic elements.

The AA, having found itself with a similar database of loyal members for roadside breakdown and other largely technical services, has for many years sought to cross-sell loans, ferry bookings, holidays, and even AA-branded mobile telephones (in alliance with a well-established mobile telephony service provider). Of course, the AA sells motor insurance and all sorts of other insurance, too. The greater proportion of the AA's profits come from its original breakdown business, but information management, effective delivery of telephone-based services, and the multiple reuse of its customer database are significant aspects of the business and will become more so.

The function of providing information to customers, generally by telecommunications-based means, is clearly a vital component of service but should not be regarded as the whole opportunity. Call reception centers of a company are not operating as independent businesses. Nor do they have to stop at being a sales channel exclusively selling their own airline's products plus some noncompeting, complementary products such as car hire. What we have seen develop is a willingness to book journeys from anywhere to anywhere—obviously seeking to use the host airline or global partners where possible. What we have also seen is the development of frequent flyer schemes/loyalty clubs with valuable customer databases. But they are not always used as well as they might be.

In the same year that I spent £15,000 ($25,000) on air travel, my main frequent flyer club annoyed me by seeking to cancel my membership on the grounds that I was no longer an attractive customer! My journeys encompassed the United Kingdom, the Far East, Australia, the Baltic States, Scandinavia, and around Central Europe, mostly in combinations of routes where, it so happened, they were weak at the time. But, having gone to the trouble of putting me onto a database, perhaps they could have served me as a travel agent for a while? However, that isn't the core business of an airline; or not yet, anyway.

But how well do *travel agencies* do? A travel agency is a service provision business. It adds value by building journeys for its customers, often selecting airlines and other components of the journey on behalf of the customer. But you are not really conscious of travel agents operating their own frequent flyer schemes (except those like American Express, detailed below, that operate as credit card companies). You tend to do business with a travel agent one journey at a time. They mostly don't seem to be doing much to use the data they have on each of us as customers.

The market model of the British rail system after privatization comprises Railtrack, which owns the infrastructure; *train operating companies* (TOCs), running passenger and freight trains; rolling stock companies, leasing trains to TOCs; and maintenance/track renewal companies, which will carry out most of the work required by Railtrack. The British government has decreed that the country's rail system will be served by a single information system—from Railtrack. Happily, BR Railtrack has now issued a request for proposals to improve an initially chaotic situation. A call center will result; will the solution go as far as building a customer database?

Surely, there will be scope for rail to go far beyond providing efficient mechanisms to react to incoming inquiries. The aim should be to attract even *more* calls...and to convert them into profitable long-term customer relationships. A single inquiry to any point should deliver information on the whole journey. Perhaps the entity receiving the call will then take bookings, collect money, and deliver tickets on behalf of many TOCs and overseas railways, too. Rail should at least aspire to the standards already being achieved by the airlines in providing basic information. If Railtrack builds a customer database, it is also building an opportunity to become a telecommunications-based service provider of many other products than rail journeys alone.

It has not escaped the attention of credit card companies that they can protect market share and establish new streams of revenue by the combined exploitation of the reuse of customer information. American Express is renowned for its mailing lists; it is also a travel agent, gives *points* for whatever you book, and maintains a *profile* of its regular customers to ensure they always book, say, nonsmoking aisle seats and vegetarian meals. Amex has also set up as a mobile telecommunications service provider, at this late stage of the game, but targeted on its cardholders. It is possible to obtain emergency assistance with one number. MasterCard, Visa, and others have also given close attention to cross-selling possibilities in some of the countries in which they operate. Often, they will put you in touch with a local plumber or a similar service.

Like many people, I have a gold credit card that is associated with my bank account. Having paid my membership fee, I am entitled to a number of valuable services including travel insurance, emergency assistance, and cash withdrawal. I gain *air miles* from my transactions. Elsewhere in my banking relationship, I handle such activities as cash management and insurance. I could also use my bank for stock market investment, but I prefer the people with whom I already deal (over the telephone, too).

4.7 ALL KINDS OF CUSTOMERS NEEDING ALL KINDS OF SERVICE PROVIDERS

Telecommunications-based service provision is already pervading a much broader range of opportunities than the first two case studies embraced. The scope of telecommunications-based service providers should also take account, however, of the range of skill levels and expectations of typical and atypical customers:

- Jenny, a personnel consultant, asked me for advice on managing the batteries of her NEC P4 mobile phone, which would no longer hold a charge. "I've only needed to charge them twice in two years...the car connector seems to look after them fine...two dealers wanted to charge me totally different prices for replacements, or a new phone for £50..." I explained memory effect in nickel cadmium batteries and advised her to discharge them completely and start again.

- Ian, a techno-guru colleague, demonstrated to me virtual reality: *electronic shopping malls,* called up some stuff on the Internet, and rued the fact that telcos didn't understand what is happening out there.

- For myself, I admit to being put off by the sheer volume of literature, the assumption that I was already a computer enthusiast, and the complexity of the offerings that arrived from Internet service providers when I first requested their brochures. Much of the publicity material is better now, but I'm still not confident that there are many Internet service providers dedicated to serving people like me.

- Steve, an American music publisher living in London, told me how happy he was with the price and service that he gets from his long-distance reseller.

- Elizabeth, aged six, has no awe of a telephone that works anywhere as she sees it. We discuss the concept of voice messaging, and she can play a Disney video in a VCR. She may never own a personal computer that is not also a videophone. Touch dialing may only be a fallback option in the voice-activated phones that her generation will have. She may never walk into a branch bank. We can have little idea who or what her information service suppliers will be when she becomes an active customer in a few years time.

- A friend and I wondered how successful a *telecommunications-based butler service* would be. There could be a single agency that would be an alter ego/guardian angel and provide intelligent, proactive service across a range of activity, all over the telephone. The human equivalents—a secretary, estate manager—are too expensive for most situations. We will be unusual if we have a stay-at-home partner who can do these things for us. These unassisted humans will lack, too, the sophisticated information management backup, access to preferential pricing, knowledge of alternatives, and so on. One call to 1-800-JEEVES might get suits cleaned, lawns mowed, cinema tickets booked, telephone bills checked, and new direct debits activated. JEEVES may arrange service with the share price service described in Section 3.6. You would be called back if your AnyCorp shares go above $30. JEEVES might set up social engagements—and "dates" too.

There needs to be a service provider to look after Jenny and the issues of the unkind jungle some of the mobile telephony industry has let itself become. We need a service provider for Ian to provide access to the Internet...so he can do whatever and go wherever; but do the rest of us have the skills and inclination to move into these realms without help? Steve's service provider is giving him a low-cost efficient service for a very simple need; but since it is essentially an arbitraging service, does it have an exit strategy or a transformation strategy for when the arbitrage window closes? Do we know what kind of service provider Elizabeth will need? Technology is moving so fast that we should probably not bother to focus that far ahead. There is enough to see just by looking around at the changing nature of banking, insurance, travel services, shopping, credit cards, and many more. For myself, I just want a service provider that understands my business needs and takes away from me the hard work of understanding how its product works.

Given the ever-reducing cost and accessibility of technology, it is becoming increasingly more economical to create and market products for smaller and smaller markets. Accordingly, we should soon reach the point where we get the products and services that we really want rather than those based upon undeveloped technology and created for someone else's perception of the average user. The relevance to telecommunications-based service provision of the *micromarketing concept* is examined in detail in Chapter 8.

4.8 SCOPE OF TELECOMMUNICATIONS-BASED SERVICE PROVISION

The previous section gets us thinking about the potential scope of service provision. What falls within a definition of a telecommunications-based service provider and what falls without? In Table 4.1, some of the enterprises can be categorized beyond doubt as telecommunications-based service providers, and there is a further selection of other types of organizations that have similarities of function and that face similar issues and opportunities.

4.9 USING TELECOMMUNICATIONS TO LINK CUSTOMERS TO PRODUCTS

Practically any business whose competitive proposition is already based around information is a potential telecommunications-based service provider. So, too, is any enterprise that would benefit from reconceptualizing its business to such a paradigm.

Table 4.1
Types, Roles, and Examples of Service Providers

Type of Enterprise	Mediation: Between Customers and...	Examples
Mobile service provider, cellular network airtime reseller (MSP) (1)*	Cellular networks, voice messaging equipment, handset manufacturers	*USA*: Motorola, Ameritech, *Australia*: Optus *UK:* Talkland, Martin Dawes, Astec, Cellcom, Vodac, People's Phone
Internet access/service provider (ISP) (2)	World Wide Web, database owners, mail boxes, transport devices	CompuServe, Demon, Unipalm, America OnLine
Long distance telephony (including Callback) (2)	PTO networks, fractional private circuits, international telephone accounting rates	ACC, Worldcom, Premiere Worldlink
Telephone banking, building societies (savings & loan) (1/3)	Cash, loans, other services of the bank and of other suppliers such as insurance companies and stockbrokers	Citibank, National Australia Group, First Direct, Natwest Primeline
Airline and other travel related reservation services (1/3)	Flights, rental cars, rail journeys, coaches and buses, accommodation, holidays, excursions, insurance	British Airways, Avis, Thomas Cook, Saga
Credit card services (1)	Merchants, money...plus shopping and an increasing range of other services that can be sold to the database of customers	American Express, MasterCard, Visa, affinity groups who brand these cards (such as trade unions, clubs, professional institutes)
Telephone or Internet shopping (1)	Products of all kinds	Delivery services
Roadside and other emergency assistance (1)	Mechanics, breakdown trucks, medical resources, repatriation resources	USA's AAA, Europ Assistance, Cigna Travel Insurance, UK's Automobile Association, and RAC
Fast food (3)	Ready meals, local delivery	Pizza Hut
Theater and other similar establishments and their agents (1)	Concerts, plays, films	First call

Type of Enterprise	Mediation: Between Customers and...	Examples
Specialist "call centers" (1/2/3)	Products and services of a third party enterprise—especially in context of a marketing "campaign"	U.S. and worldwide: Perot Systems, UK's Programmes (also debt collection services)
In-house call centers and any other telecommunications-based customer facing department of an enterprise (3)	Products and services of the enterprise	Sales, service, repair department

*Figures in brackets refer to definitions to be explored in Section 4.12.

This book is concerned with the bringing of service to customers using telecommunications links, rather than face-to-face interaction. Both *information technology* and *telecommunications* go some way to define the nature of the inputs, processes, and outputs of the businesses being described. *Information services* is already well understood as a definition that covers many enterprises from Reuters to libraries. The focus of this book is upon the developments now arising from telecommunications and therefore the term *telecommunications-based service provision* is preferred.

Telecommunications-based service provision is not just about voice but also about data, video, and other elements from the explosion of information technology. Telecommunications-based service provision may include carriage, switching, and the supply and support of relevant or associated services and devices (e.g., cellular service provision and international resale, handset supply). Computer telephony integration and back-office computing aspects are largely opaque to the customers where the interface is by voice; it is the telecommunications link with the agent that is being experienced, whether that agent is human or a voice processing machine. There is anyway a strong case for putting computer telephony integration at the center of the piece.

Voice may be the primary interaction medium for many businesses, but the computer keyboard (or mouse, etc.) are often just as relevant, especially with respect to hard-wired or dial-up information services and the Internet.

The Internet can, in any case, be used for voice interactions, if only for an imperfect form of *toll-skipping*. The keyboard is not the only interface in use for access and manipulation of data, even in multiple media. Interactive voice response systems provide voice-text-voice protocol conversion. Already possible is small, medium, and large vocabulary voice recognition, some natural language capability, and some interaction with network browsers to reach home pages and with virtual reality agents in electronic shopping malls.

But the detail of the technology is not the defining issue. Service provision is defined by the *what*, not the *how*. Technology, especially convergence of telecommunications and computing, is only part of the overall story. Chapter 1 examined convergence of enterprises and other factors, too. Service providers link customers to the substance of entertainment, education, culture, shops, and information databases of other kinds. Figure 4.1 indicates that the information market place comprises two main classes of player: owners of content and gatekeepers of customers.

4.10 POSITIONING AND CHARACTERISTICS OF SERVICE PROVISION

It is the diagram in Figure 4.1, perhaps better than any other, that illustrates the role to be played by telecommunications-based service provision. The following definitions and characteristics can be considered:

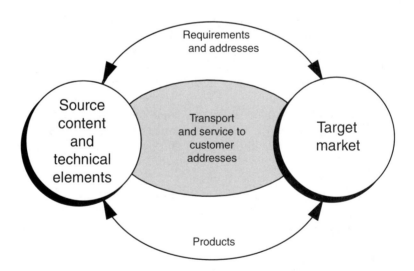

Figure 4.1 Source/content, channel, and customer.

- Telecommunications-based service provision links selected customers to the owners of information content.
- T-BSP creates services that comprise information, technology, and other elements and presents them to customers in an easily used manner.
- Service providers will normally act on behalf of their customers in selecting the suppliers of the elements.
- Service providers act in alliance with their suppliers, but will be free to change suppliers if it is beneficial to do so.
- Service providers will normally be more than a sales channel for a single supplier—but such channels will have many of the characteristics of service provision.

In telecommunications, many pricing issues and commercial relationships continue to be regulated. Service providers, to a greater or lesser extent, can help customers iron out issues that this presents.

The value that is added by a service provider may include some or all of the following: knowledge of customer requirements, knowledge of elements to be used, buying power, ease of use, time savings, personal attention, future-proofing, high levels of availability/reliability, consolidated billing, and service in all required geographies and at all required times.

4.11 BOUNDARY OF RESPONSIBILITY

Generally speaking, we assume the service provision processes to be separate from the construction, development, or operation of the technologies themselves. However, a good level of technical understanding is needed in order to assume the role of making technological decisions on behalf of customers and to select and integrate the best of what is on offer.

Furthermore, if the service provider takes responsibility for quality of service (as surely it should) then it requires the ability to challenge and work with what is being presented as fact by the element providers (such as telcosand software designers).

4.12 DEFINITION OF TELECOMMUNICATIONS-BASED SERVICE PROVIDER

Building upon these characteristics and the value to be added, three generic forms of service provider are possible:

- In the *first form*, the SP buys information and technological elements from suppliers and resells them to end customers, adding value for the customer such as tariffing, billing, or equipment supply and for the supplier such as marketing, channels, billing...(e.g., cellular service provider, video hire shop, and credit card company cross-selling elements from a third party).
- In the *second form*, the SP also *switches* or otherwise processes the information or technological elements before reselling them (e.g., Internet service provider, international/ long distance telephony resellers), adding value for each.
- In the first two forms, the assumption is that the SP and the element providers are independent of/external to each other and that their relationship is defined by that factor. A *third form* of service provider is an *in-house service provider* (e.g., customer sales and service helpline, airline reservations line).

The examples in Table 4.1 have been classified according to this model. There are hybrids, too.

4.13   VALUE CHAIN OF TELECOMMUNICATIONS-BASED SERVICE PROVIDER

All three types of service provider operate the same overall processes.

All enterprises can be characterized as a value chain. The value chain is the starting point in describing processes and their interaction, as shown in Figure 4.2. Enterprises have inputs from suppliers. They do things to the inputs, then supply the outputs to customers. The difference in the value of the outputs and the costs of acquiring and processing the inputs provides a margin to acquire and maintain working capital, reward stakeholders, and pay taxes.

Delivering a competitive *service provision* proposition requires a sound understanding and practical application of the following process areas: creating products, obtaining orders, implementing orders, and delivering continuing service. These primary activities need to be thoroughly integrated and supported by the organization; there must be continuous communication to align the interests and expectations of customers, suppliers, employees, and funds providers.

4.14   WHY TELECOMMUNICATIONS-BASED SERVICE PROVISION MAY BE RELEVANT TO YOU

Given the advances in information management and telecommunications, in terms of cost and of functionality, barriers to entry are falling

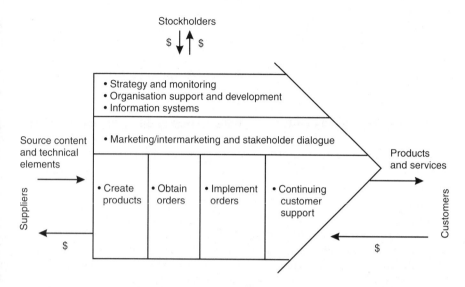

Figure 4.2 Value chain of a telecommunications-based service provider.

away. Even if a business conceives itself as an "airline" or as a "bank," or even as a "restaurant," it behooves the managers of that enterprise to look at ts business again in the domain of telecommunications-based service provider if it is to avoid the risk of being marginalized or finding itself competing in a commoditized market. Telcos are as vulnerable as anyone in this regard.

A few years ago a telco executive asked a colleague in Reuters how they made money. "That's easy, she replied: You charge 50 cents to carry the message; we bring together valuable content and charge $50 to a trader who makes $5,000 from the deal." Admittedly, the telco carries millions and millions of messages and makes a lot of 50 cents, but the price of carriage of signal is trending steadily downwards towards the cost. The Internet is virtually free already. Even though the volume of signals will increase beyond measure, many telcos will go out of business under the weight of costs and investment in their current paradigm. Many jobs will be lost.

In late 1995, the *Economist* produced a study of telecommunications entitled "The Death of Distance." Banks are shedding staff in their branches. The property value of those branches will decline. We are probably set to see yet another dramatic change in the landscape of our town centers. The work is transferring to call centers in this country and overseas.

We have already seen great changes in our lives and our businesses over the past few years resulting from advances in telecommunications and computing. We have seen only the merest start of business and social interaction being transacted over the Internet. This poses both threats and opportunities.

Service providers who can select and address a market, manage technology, take advantage of convergence, and manage suppliers and subcontractors will be well placed to provide a real value proposition to customers that will make them indispensable in their businesses or private lives. The quality of a customer database may be the key determinant of relative value and competitive opportunity. With a good customer database and the entrepreneurial flair to exploit it, a telecommunications-based service provider can tilt the balance of power away from the producers of basic products. The prizes will go to those who can create and maintain a valuable and sustainable information service proposition. Processing, storage, and carriage have been commoditized; information as a product is of no value until it has been shared and paid for.

Given what is already apparent to us and what is still to come, any business that is information-based in any conceivable way needs to consider its strategic position and its strategic options in the light of telecommunications-based service provision. Chapter 5 is the other half of the conceptual analysis of telecommunications-based service provision. Part Three follows with a systematic description of how telecommunications-based service provision is carried out in practice.

Organization Design 5

5.1 INTRODUCTION

The general processes described in the value chain at the end of Chapter 4 will be commonly found in any buy-work-sell enterprise. The particular ways that these process areas operate in a telecommunications-based service provider are developed and detailed throughout the book.

There are many characteristics of a successful business that are also common. Service providers should concentrate on these six:

- Strong sense of direction to guide it through change;
- Sound competitive proposition that is well-understood;
- Speed in taking products to market;
- Skilled, committed people empowered to work for their customers;
- Systematic but flexible processes and tools that save time;
- Superior teamwork. Clear roles and responsibilities.

The body of the chapter is divided into five parts:

- Development of organization design over time;
- Socio-technical systems—people and machines;
- Processes;
- Working together;
- Pulling it all together.

5.2 DEVELOPMENT OF ORGANIZATION DESIGN OVER TIME

5.2.1 Organization Design in the Telecommunications Age—Major Paradigm Shift

What has happened to our paradigms of enterprise organization? How do we identify the requirements for information-based organizations in the new telecommunications-based age? What approaches an "ideal" organization architecture, upon which a telecommunications-based service enterprise can be built?

> "The ecology of our world business environment is changing dramatically. Gone are the days of the comfortable protected market niches. Gone are the days of geographic isolation.... Change in technology is also everywhere.... Nowhere is change going to be more dramatic than in the way we manage and lead our enterprises. The logic of computers and networking makes obsolete many of the deeply cherished notions of the past.... Just as long distance runners sometimes "hit the wall" of their capabilities and endurance, many companies are hitting the organizational wall, or so it seems. They have trouble absorbing more computer-based technology. The reasons for this are often unclear. Could it be that we are putting fifth-generation technology in second-generation organizations (Savage 1990)?"

Eighteen of the 36 enterprises lauded in one of the defining books on management, *In Search of Excellence* by Peters in the 1980s, have since failed. In how many of those now failed enterprises was one of the causes of failure the way the enterprise was organized, its processes, structures, and systems? Is it that what worked then may not work now?

A system will not survive if it responds to external changes more slowly than those changes are taking place. The old organizational paradigms don't work in the new knowledge-based age. Many of the organizational improvement initiatives, such as *total quality management* or *putting customers first* have not worked, or have failed to deliver what had been hoped.

Savage illustrates the development of organization through a series of historical eras in Figure 5.1.

The earlier organizational designs and methods worked well in slow-moving, predictable, relatively simple environments with limited customer choices. But a new paradigm has become necessary for those enterprises that have to accommodate simultaneous situations of rapid

Historical Eras

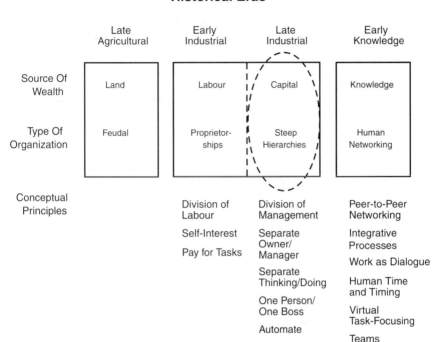

Figure 5.1 Organizational form in successive eras (Savage).

change, confusion, and demanding and wealthy customers shopping in a global and instantaneous market.

Control of enterprises was formerly vested in a few people at the top. Now a lot of the power has shifted; nowadays, much more of the controls in an organization have to be exercised at the point of delivery of services to customers.

Globalization, deregulation, and the arrival of the telecommunications age have caused the new environment. We have, on the whole, not yet worked out our organizations to cope.

5.2.2 Cooperative Paradigm as an Organization Structure

Ericsson approaches the development of organizations over time as the move from whale to scattered fish to shoal.

The whale represents a formal, hierarchical structure. It is "big, coordinated, efficient but slow at turning round and adapting to changing environments." Also, when the water becomes too shallow, the whale is beached and it dies.

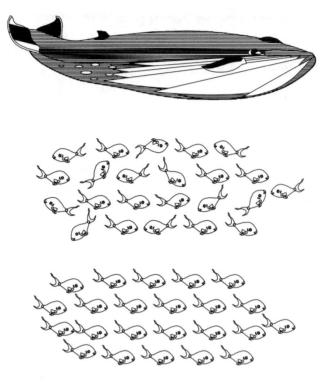

Figure 5.2 Organization development from whale to shoal structure.

Major telcos, banks, and computer companies all have or have had these characteristics at some time—though not all of them have been beached yet.

When organizations are broken up into profit centers, they "behave like scattered fish, swimming in whatever direction they please." They are good at spotting and taking opportunities, but they are vulnerable to being eaten by a bigger fish (with a more efficient cost structure or some smart alliances).

So scattered fish come back together into shoals:

"Flexible and coordinated at the same time. The shoal combines centralization and decentralization—autonomous units are held together by a common destination and by common movements. Simultaneous movements make the shoal appear like a coordinated whole—local and global actions for the same cause...in business, shoals are held together in virtual integration by a common purpose, a common vision (view of the world and values), and a common IT."

The value chain in Chapter 4 may look rigid and as if it does reflect the spirit of the "shoals of fish." However, it is much more so when you add in the information links and when you populate it with helpful, intelligent, and adaptive "people as computers and communicators."

Information technology has provided the means whereby organizations can be built to share information and, assuming less need to be "in control" and "driving," the leaders of organizations can concentrate on the processes of articulating direction and values, motivating performance, procuring resources, and helping groups to adapt and interact.

5.3 SOCIO-TECHNICAL SYSTEMS—PEOPLE AND MACHINES

5.3.1 Organizations Are a Combination of Social and Technical Systems

In primitive societies, it was said to take a team of 11 people to find, chase down, corner, kill, skin, dismember, and carry home a large wild animal. Some members of the team were good at tracking, some good at running, some good at throwing, and so forth. Tools were rudimentary.

As the industrial revolution got into its stride, however, it was tools and machinery that drove much of the successful evolution of industry and businesses—though not exclusively.

Adam Smith propounded the advantages of specialization (i.e., "the pin factory"). Frederick Winslow Taylor showed people the "one best way" of doing things, sometimes portrayed as having one set of people doing the thinking and another set of people doing the work. The next generation of management thinkers, especially in the Hawthorne experiments, took, once more, greater account of the effects of human issues and group dynamics. Study of work followed the fortunes of Henry Ford and, much later, the Volvo experiments whereby people were taken back off the assembly line and set to produce in self-managing groups.

It became widely accepted that social and technical considerations were present in the performance or nonperformance of an organization and that the two aspects interacted. The term *socio-technical systems* was coined.

The golden peaks of informatics mechanization have brought us untold computing power to solve linear programming problems, to analyze multiples of variables against other equations, and to resolve multiple resource planning issues. Drawing offices have tools to describe and record. We have better ways of creating text. We have the capability to do more better and cheaper, and that capability will continue to grow.

We have better technical means to communicate than ever before. It is superior performance in communication, in particular, that is setting apart some enterprises and putting them ahead of the competition. Rapid and successive developments in technology, especially the remarkable acceleration of information transfer capability, have brought in their wake remarkable new business and competitive possibilities.

But these explosions of information, possibility, and change have made it more and more difficult for individuals to perform effectively in organizations. In an attempt to exert more control, management structures became larger and more complex as organizations went through the giant phase of the late industrial era. Dysfunctions of process and conflicts were resolved by referring to the person at the top of the organizational pyramid or by the creation of liaison functions. It doesn't work in the fast-moving telecommunications-based service provision environment. In the face of excessive change in the environment but inflexibility in our organizational response, we suffer what Alvin Toffler has referred to as *adaptive breakdown* in ourselves and, by extension, in our social groups.

An enterprise is a socio-technical system. The *technical* is ever stronger, but there have been new rules to learn in strengthening the *socio*. If the latter becomes weaker, then the product of the two will remain the same or even decrease. Given its competitive and fast-changing environment, the telecommunications-based service provider needs to be powerful in both respects.

5.3.2 The Human Being Must Now Be Valued Above the Machine

In the early equations of the socio-technical system it was the machinery that was expensive and the labor that was plentiful, cheap, and compliant. We grouped ourselves around the machines, and we worked at the pace of those machines, as depicted in Figure 5.3.

When people were valued far less than the machine was valued, it was people that paid the price for greater economic efficiency. They suffered death and maiming. There are many places in the world where this is still the case today. But what happens when a person is in physical danger in a modern society?

No price in helicopters, fire engines, other peoples labor, or material is put above the price we will pay to save life or limb. For example, too, today's cars are built to collapse gracefully when they crash, dissipating the unwanted energy away from the frail and precious occupants.

The same value principles have come to guide us in the way we support our humans in their work. Machinery has become relatively cheaper and the people relatively more precious. The machines are more powerful

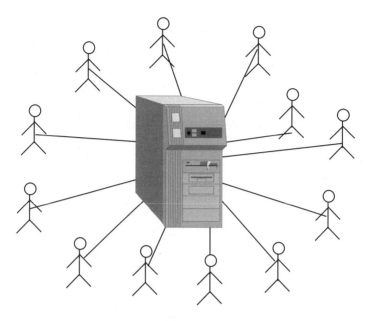

Figure 5.3 Humans grouped around the machine they serve.

now and work faster than before. Each of us has access to powerful tools that release us from drudgery and provide us with margins of time, since the time is no longer all spent on mere survival. In the workplace, we are served by elevators and air conditioning and plentiful light. No modern farmer is without power cultivation and harvesting equipment. Few modern households are without television, washing machine, power lawnmower, and one or more cars. The cars are made easy to drive; they have many built-in comforts. So each of us has access to untold and largely redundant functionality in the machines that serve us. We also have wide access to the world's store of knowledge, information, and data.

The model of the people grouped around the machine has been redrawn as machines grouped around each person, as shown in Figure 5.4. Information is taking over as the dominant factor of production. Since the costs of processing, transporting, and storing information are tending quickly to the nearly free, so the relative costs of the *machine* and the human being have now changed fundamentally in favor of the latter. A modern organization values its precious and expansive humans above its ever-cheaper machines.

Note that the examples and discussion above are centered on the physical well-being and comfort. Many people think that society does recognize the value of emotional well-being in the workplace and its effect on productive capability. Society's response is often thought to be

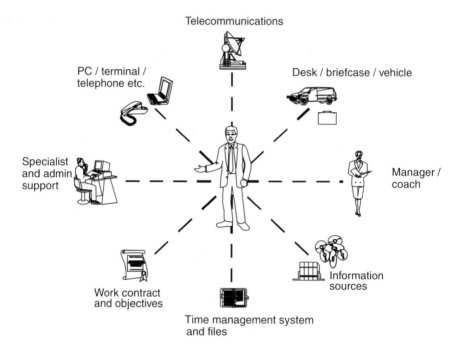

Figure 5.4 Human being at the center of the socio-technical system.

largely charitable in responding to emotional breakdown, rather than enlightened, just, and energetic in designing organizations and social dynamics to prevent it in the first place.

5.3.3 People Limited by Their Organizations?

If the potential for the *techno* is virtually unlimited, then is it now the *socio* that is the limit of our effectiveness to perform?

A delegate at a World Congress on project management spoke as follows:

> "We have all the hardware, the technical systems well developed. We are comfortable controlling technology, money and suchlike. But it's the software, the programming of the team...we don't think about how to program the human system to perform as we want it to; our attention goes on the technical aspects, and yet in our heart of hearts we know that they don't go right unless you also get the people factors right."

The issue of "people" goes much deeper than questions about their commitment and skill.

A 1970s study of management asked why these truculent, useless people were, at the same time, so effective when they were *outside* work. What caused these people to be so unsuccessful at work and yet so positive and powerful as amateur footballers, musicians, chess players, and leaders of Girl Guide companies? Could it be the way we set up the dynamics of our organizations?

Whereas formerly we had access to small amounts of slowly changing information, we are now subject to greater and greater amounts of information arriving at shorter and shorter intervals. We have greater expectations of ourselves because we see that all this power is there at our fingertips. At the same time, people in many information-based enterprises report feelings of greater power*less*ness. In particular, people report not being able to provide good services for their clients. Here is the heart of the problem. It is not the fundamentals of technology in our products that are at fault. With exceptions, there is as much technology as we want. It is not us as the people in our organizations who are at fault. With exceptions, we are well-educated, hardworking, and cooperative.

5.3.4 Humans as Computers and Communicators

Many organizational leaders say things like: "It is our people who are our greatest resource." This often seems to imply a valuing of people in an emotional sort of way. We recognize that the variations of performance of people *as social beings* do make a world of difference to the perceived delivery of services. We generally try to be nice to the people in our organizations and they enjoy the social side of working with us. But does it make us that much more effective? We do perform better when we are protected from excessive stress. But aren't many of our organizations seriously *under*valuing two of the primary skills of the human being? "Information" has arrived on the scene as the fourth factor of production. What is the human *in the domain of information processing and communication?*

Effective information processing has become a potent factor in the relative performance of organizations. We invest large sums in our computers and telephone systems. But, ironically, in conditions of high complexity, multiple inputs, conflicting values, and rapid change, *it is often the human who accesses, stores, processes, and retrieves information best.* Perhaps we can call some of this "gut feel." Certainly there is much reliance put on just that in many organizations today. Many organizations, too, value the very differences in the ways in which different humans interact with information. Some are best at creatively connecting one idea with another; some are excellent at mental arithmetic; some can process in more than one language; some can program computers; some can "remember" better than others, and each of us brings different facts and emotions and experiences to the remembering process. In the information age, therefore, an enterprise will do better when it has created an environment and provided tools for the human computer to do better.

Given the right environment of calm and safety and comfort, the human computer will get on with it's work. No computer can work in isolation; inanimate and animate computers alike must communicate with other computers. *Effectiveness in communications, especially telecommunications, will be key a determinant of business success, often providing more leverage than the underlying computing or other business processes themselves.* Similarly, it may be the effectiveness of the communication between the "humans as computers" that will determine levels of success.

From our earliest life, we learn to communicate with those close around us. First we communicate with our mothers, then our father, siblings, and other close relations and neighbors. The recent work on *emotional intelligence* is showing us the paramount importance of working with our emotions and upon our communication with those around us. Girls are generally thought to have more emotional intelligence than boys. It is not the remit of this book to discuss gender factors in the workplace or anywhere else, but there is plenty of evidence to suggest that men's ways of doing things are related more to power than to cooperation. There are also studies that suggest that tomorrow's managers will need to deploy more skills from the "feminine repertoire" than before. Emotional intelligence, including effectiveness in communication, is being portrayed as the prime determinant of "success in life."

5.3.5 Working in Small Groups

Our ability to communicate inside small groups is determined long before we come to work. These basic groups are probably around five to ten people. The members of those groups, especially the leaders, do need to satisfy themselves that new and existing inhabitants of those groups have the basic skills and tools for their job and that they are sufficiently motivated and adapted to work successfully in achieving the tasks that are set for the group.

John Adair (1960) says that leader of a group is there to ensure that it:

- Achieves the task;
- Builds the team;

- Grows the individual.

But neither the greater ability of small groups nor their key role as building blocks of the organization are recognized as often as they should be. Or, else, these groups are compromised because they are not given power and information. Adair's defined *tasks* assume direction is given from elsewhere.

5.4 PROCESSES

5.4.1 Time as a Competitive Issue

Alvin Toffler talks of slow nations and of fast nations. Those that have an effective postal, telecommunications, banking, customs, and administrative infrastructure will be considerably more competitive than those that do not.

Time to market for new products and "time to customer" (how quickly a customer can be handled effectively when calling) are key determinants.

What advantages there are for an enterprise that has high levels of contact with customers and a capacity for fast cycle action.

In their book, *The Virtual Corporation,* Davidow and Malone enjoin us to greater contact with customers, since, as they put it: "The ideal (virtual) product or service is one that is produced instantaneously and customized in response to customer demand."

Organizations should, on the whole, be designed around their *time base* rather than their *cost base,* for it is time that often determines revenue and gross margin. Consider the following examples.

- A representative from a leading bank told a customer that it would take "four to five weeks" from starting the loan process to receiving an offer of mortgage. A telephone banking and loan organization offers to do it inside 20 minutes. The internal features of each of the mortgages on offer is likely to be the same.
- Ericsson has two diagrams that show a "value package." Consider the increasing value of a courier service as its time to deliver decreases. Look how the value goes up as the product proposition is moved to a place further inside the value chain of the customer. "Floor to floor" takes up less of the customer's time than "door to door."
- A better thought-out dialogue in an IVR system will reduce the occasions that callers abandon *self-service* and opt to be transferred to an

operator. Proper trimming and a shorter dialogue reduce the time that the caller is on the phone. These improvements reduce at least three things:

1. The time that customers have to invest;
2. Operator time, therefore operator numbers can be reduced and costs saved;
3. Call toll time.

Figure 5.5 shows a practical example.

5.4.2 The Nature of Processes

Aspiring to a "factor of ten or even one hundred" improvement takes the brakes off the imagination when it comes to designing process improvements.

The *overall function* or *overall process* of an enterprise is that it takes inputs, principally from its suppliers or the environment at large,

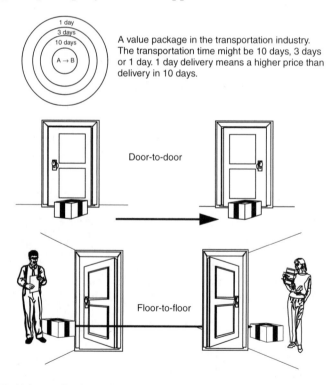

A value package in the transportation industry. The transportation time might be 10 days, 3 days or 1 day. 1 day delivery means a higher price than delivery in 10 days.

Door-to-door

Floor-to-floor

Figure 5.5 Value packaging—time and place (Ericsson).

and acts upon those inputs. It thus adds value and provides products to its customers:

input→action→output

An enterprise such as a telecommunications-based service provider will typically have approximately eight "process areas" containing 40 or 50 main processes. All of these processes are interlinked at some point and all are necessary for the achievement of the whole. Often, the outputs from more than one process provide the inputs to one or more other process. Inputs and outputs often link back into prior processes, both as feedback loops and as subprocess strings (e.g., a minor, but essential, process to print an address label for a customer may loop right back into the original subprocess of "capture address data" within the process "handle initial sales inquiry"):

input→action→output/input→action→output/input→action→output

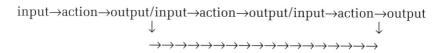

Process areas include *activities* that may not be critical or may be too unimportant to document in detail (e.g., the simple activity of allocating a pool car to an individual for a day may not be deemed to have sufficient information value to be included in the process of *maintain vehicle record* files). Other activities may be essential or desirable for the safe and effective running of the enterprise but do not have an immediate impact on the main value chain of the enterprise (e.g., issuing passes to visitors, watering the plants, washing sales employees' cars). Nonetheless, the failure to carry out such activities may leave customers or potential customers with the fear that critical processes are also handled in a slapdash way.

Processes and activities in an information-based enterprise will be linked together through the supporting information technology, or, rather, they should be.

Processes and process areas can, and should, be documented. Each process area for the telecommunications-based service provider is described in Chapters 6 to 13. In each of those chapters there is a table that shows the following information:

- Process area name;
- Owner, other actors;
- Function;

- Time to complete;
- Inputs (and sources);
- Outputs (and destinations);
- Support documents;
- Effectiveness measurement (measured by those receiving the outputs);
- Risk and cost management/disaster recovery.

Processes and their attributes can be flow charted. An example, "Conduct Major Bid," is given in Chapter 9 (especially note Figure 9.5).

Does the entrepreneur sit down and define the processes of the enterprise before starting? No. What the entrepreneur does do, however, is think through the things that have to be done and build a metal picture of the process map and of each crucial process. Just because a process does not appear to be formalized or even particularly apparent, it does not mean it is not being carried out. Any reasonably successful organization coming to document its processes for the first time may observe that it does not do some things; logic dictates that it *must* be doing them, however sketchily, to be in business at all.

"You should always do an account development plan—that is, a plan that lists the forthcoming business requirements of a major customer, describes how you intend to interest them in our company, and what products you intend to promote."

"We don't do account development plans."

"You may not think you do. I agree they aren't written down. However, I heard your boss and you agree in the corridor that X Corp. needs to improve its telephone answering and that you'd both go and see its customer service director next week to propose the use of interactive voice processing equipment."

"Yes, I did. Come to think of it, although I am very stretched for time, I also made some notes for the meeting...and I started a file, too. Is that an account development plan?"

What about the quality manual prepared for ISO 9000 registration? On the whole, processes are well documented. However, what may be missing are the real ways that things are done; they may date from the last inspection and not be amended until the next. If you really want to know how things work, then go and ask Fred.

It's in Fred's Head

After 10 years with the London Underground Railway, Fred left his job in the workshop refurbishing brake assemblies. Two weeks later started a spate of failures—trains began to overrun the stations at which they were pulling up. The brake assemblies of the offending trains were checked and found to meet all of the laid down specifications. Soon somebody thought of asking Fred about the problem. During his apprenticeship he'd been shown how to do the job..."Fill the fluid reservoir up to here..." (somewhat below the mark). The book process, followed religiously by Fred's successor, called for a top-up right to the mark. This caused a spillage on to the brake lining showing up once more an original design fault of 20 years before. London Underground then obtained Fred's help in rewriting the procedure and the problem went away.

So, are process documents actually worth having? Are all our main processes the same in character? Do they all need documenting the same? Will people read all this documentation, even if it does get written?

5.4.3 Rectangles and Fuzzy Stars

Take a phone company. It transports bits from one place to another through a complex system of switches working to the software instructions of routing tables. A friend of mine, Dennis, worked in an Australian telco looking for bugs in the software that were causing misrouting problems. Even one misroute in a thousand was thought to be a problem, for who knew what might happen or what knock-on there might be if a more serious or complex software condition arose. A few years ago, most of the telephone system of the northeastern United States was tumbled down for just such an obscure reason.

Some things in a system have to be utterly predictable. Failure in some systems causes knock-on elsewhere. Some systems have the same things happening through them again and again and again. The examples above are technical. There are socio systems that operate in the same way. Precision, predictability, impartiality, and consistency are valued above all else. I feel this way about the work of the clerk, with her computerized tools, who calculates my monthly pay. She follows rules and written procedures unflinchingly; she knows the applications and operates her computer diligently and undistracted.

I call these processes *rectangular*. They are regular; they have straight edges that interface precisely with other rectangular processes. When a rectangular process has to be changed, it must be changed into another rectangle. Rectangular processes cannot be changed whilst they are in operation. They must be isolated or the entire system shut down whilst changes are made. The resulting changes must be tested before the system goes back into service. The operation of, and occasional alterations to, a telephone network is a prime example. Other examples could include a loan interest algorithm in a banking application, the development of a camera film, or an aircraft movement logging process in an air traffic control system.

For a *rectangular* process it is essential that the process is understood and documented to the finest degree of detail; it is essential that the information systems and the people operate to precise rules; there is no room for variation. Rectangular processes work fine for predictable inputs and unchanging environments. Rectangular processes by their nature tend to handle large numbers of transactions, but many aspects of market planning are also rectangular in nature. Accountancy is rectangular—applying known rules and recording practices to largely repeatable circumstances.

But many processes are less structured. They are subject to unpredictable inputs. Many have complex conditions and variations in required output. So, the processing rules are difficult to write. I call these *fuzzy star* processes. More is left to the sensitivity, intellect, judgment, and skill of the process operator. Thoroughness and attention to detail may be less valuable in these circumstances than creativity or ability to detect patterns or to detect weak signals. The deployment of listening and questioning skills, explanations, and attention to nonverbal clues may give a higher payoff than an unequivocal request for input information or a simple transmission of structured data. Examples of fuzzy star processes include business development, some aspects of the sales process, public relations, brainstorming, and piloting a helicopter without a computer-aided flying control system.

Some processes are fuzzy when they should be rectangular: When the 984 radar system was linked to the new Action Data Automation system on the aircraft carrier Victorious in the 1960s, it was expected that the control of fighter interceptions would be largely automated. Speed, direction, and height data of both fighter and intruder would be used to calculate vectors for the perfect interception. Alas, data from the system was presented 20 times per minute but the radar sensor could not discriminate better than 2 degrees of azimuth or 200m of range. This margin of error provided variations of input to the precisely accurate computer that produced instantaneous and magnified oscillations about in the true answer. It took an experienced human to remove the nonsense before passing instructions to the fighter. Fighter interception remained an art rather than a science for a good few years thereafter.

Some processes are rectangular when they should be fuzzy. This is typified by the *jobsworth* syndrome, whereby an employee applies an inappropriate rule to a situation instead of common sense. "More than my job's worth ma'am." Some rectangular processes are given arbitrary fuzziness or leeway to prevent their overfrequent exercise—for example, the traffic warden who will normally wait a while before writing up a parking ticket.

In some tasks, either characteristic may appear at random. Customer service and complaint handling are prime examples. Often these jobs are tough to do. More fun, but often tough, are horseback riding and the sailing of boats—there are rules to follow and there are expectations of outcomes but wind, waves, and horse logic are not always predictable. Skiing is normally rectangular or largely rectangular. *Design* is an interesting mixture of the two. Tax planning starts off creative and fuzzy and then becomes rules-based and rectangular. Child care may be the opposite. Many of us find computing fuzzy. Many of us handle the VCR in a fuzzy way because we cannot remember the rectangular processes that dictate its use.

Finally, some processes are fuzzy on one side and rectangular on the other, for they are deliberately there as translation and conversion devices. For example, the salesperson takes in a fuzzy request from a customer and translates it into the structured forms that will provide an input to activate a product. Alternatively, and more frequently now, the customer could have a firm requirement, detailed in a request for proposal, against which we have to amend and project our product in a bid document.

A telco sales force has a fuzzy edge to talk with customers; the back end of the sales force is a straight line (rectangular), so as to convert the fuzzy input from customers into the rectangular forms that will be needed in the networks...and so forth. See Figure 5.6.

5.4.4 Process Measurement

Process owners must be "energetic champions" rather than "process guardians, indomitable protectors" (Obeng and Crainer, *Making Re-engineering Happen,* 1994).

Broadly speaking, people and small groups will perform willingly and well if they are given tasks for which they are capable, for which they are resourced, and that they believe are both necessary and appreciated. At the core of process management is process measurement, for it will provide a yardstick for each of those dimensions.

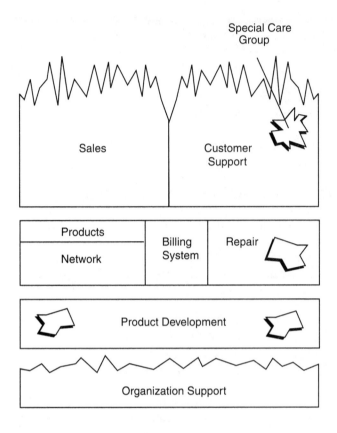

Figure 5.6   Rectangles and fuzzy stars—a telco.

Sometimes it is better to allow process owners to do their own measurements, but it is normally better to encourage measurement by the receivers of the outputs.

Dennis (mentioned earlier), who worked in the bowels of a major switch code, has also been CEO of a radio-paging company in Moscow. The market, the company, and the people were all new. So, too, were the notion of Western-style accrual accounting and of personal power and accountability for any but the highest ranks. Dennis thought to ask for performance information from each department. After some weeks of frustration and mutual incomprehension, he changed his tack. Instead, he asked each department head for any information that *they* thought would describe the essence of what had happened the week before. Soon Dennis had information on sales volumes, product mix, customer calls made, revenues collected, and bills outstanding...It was no great task for Dennis to collate the information into very serviceable management accounts. The individual managers were pleased with their own new personal yard-sticks and performance went from strength to strength.

On the other hand, BT and the Telecommunications Managers Association took a long time to agree on a method of measuring private circuit installation performance. Neither would accept the other's version of the facts. In my view, it was the telecommunications managers who were the customers and obtaining value or not from the performance and it should have been they who defined success or shortfall. It is sometimes said that no system will improve in the absence of powerful external stimuli.

In any case, measurement should reflect an alignment of all stakeholders' requirements. Process teams will normally improve what goes on inside their boxes in response to external stimuli. Those stimuli will be effective if they postulate issues in terms of customer-stated values/requests for improvement and ineffective if they are perceived as subjective and offensive criticisms of skill or commitment.

5.5 WORKING TOGETHER

5.5.1 Moving the Locus of Control Closer to the Action

If "the customer is king," as a frequently seen piece of management graffiti would have us believe, then does not the day-to-day sovereignty within the serving organization need to be put really close to the customer? Given immediate access to the knowledge that enables decisions and causes actions, should we not be concentrating on engaging employee maturity and commitment to make things happen?

It should not always be the person at the top of the hierarchy who takes charge.

Consider for example, the work of Mary Parker Follet in the 1920s, which referred to the "authority of the situation" rather than of the person. I often wonder how well she was regarded at the time. As a young officer in the Royal Navy, I received short shrift from my superiors when I suggested that in certain circumstances I put myself under the instructions of my highly experienced and specialized chief petty officer.

Logic of computers and telecommunications has certainly put aside "obsolete notions" of control from the top as the most effective organizational form (though in conditions of conflicting priority it is still the best way). In the new telecommunications-based world information upon which to make decisions is, or can be, available to all in the organization.

5.5.2 Redeeming the Tensions in the Ways We Work Together

In *Making Re-engineering Happen,* Obeng and Crainer identify a set of tensions between people, between processes, and between established organizational values and emerging organizational values. They see one part of organizational re-engineering as the conversion of tensions into new energy, as shown in Table 5.1.

5.5.3 Constituencies and White Spaces

People are largely uncomfortable with the definition of *process ownership* and the notion of being accountable for one, but not necessarily both, of underlying process effectiveness and the actual outcomes from its use.

Table 5.1
Re-engineering to Remove Dysfunctional Tensions

Tension	Action	Energy
Standard inflexible processes	Analysis, elimination, clarification	Customer-based Flexible Responsive
Formal systems	Analysis, reduction	Informal Flexible
Between new and experienced people	In-group and out-group	Teamworking, active listening, learning and unlearning skills
Between committed and mobile people	Clear communication channels Flexible Alignment of objectives	Mutual support Career development
Autocratic and participative managers	Reduce hierarchy Collective decision-making	Managerial coherence
Supportive and blaming managers	Teamworking at managerial level	Energy focused, open communication
Risk-taking and risk-aversion	Teamworking Clear statement of values Consistency	Backing for risk
Fragmented and integrated values	Statement of values Constant communication of values through actions	Direction Organizational sense of purpose and esteem

After: Obeng and Crainer, *Making Re-engineering Happen.*

For example, one person was responsible for the development of the major bid process in a telco, for its roll-out across the organization, and for subsequent improvement. But the owner of the process was not responsible for more than a few of the actual bids that were subsequently sent out.

People often find it easier to understand constituency management than process management. Consider the following scenario.

The CEO of a mobile service provider sat down with his new management team. Finding it hard to be sure of their process accountabilities, the team nonetheless found an instinctive path to understanding the external "constituencies" for which we were responsible. The financial controller was clearly responsible for our relationships with the auditors and tax authorities. He, too, would be our sole contact with the finance people of the holding company. The marketing manager would handle the press (or supervise others, including CEO, who would do so on her behalf). She also managed relationships with product suppliers—networks and handset suppliers. Having established, with little difficulty, which of them would handle these relationships, it was that much easier to assign accountability for processes: The marketing manager would protect visual image, since only she would authorize printed material going out of the company; she would also provide the designs for building signage and for the refurbished reception. Finance might draft our customer contract forms but the marketing manager's team, also, would have to sign them off before they went into the public domain. With accountabilities of specific tasks assigned, we were able to make improvements, lose the feeling that everything we did was a crisis on which we all had to meet and agree, instead of being a matter of simple prewired process with the parameters adjusted by the person on the spot. The CEO was also largely able to step back from detailed specialist areas and operational matters and concentrate on strategic direction, setting objectives, monitoring performance, encouragement, and motivation.

The laying down of constituency management and the defining of roles and responsibilities goes a long way to ensuring that there is only one person accountable for each process and that there are overlaps or even *white spaces* where there is no assigned manager.

5.6 PULLING IT ALL TOGETHER

5.6.1 Implications for the Design of the Service Provider Organization

For all of the inborn or already developed abilities of the human to *compute* and to work with others in small groups, the enterprise as a whole will come to nought if the group is ineffectively connected to other groups and the information that those other groups have. Similarly, each group needs to have its built-in will to achieve harnessed and aligned with those of other groups.

The task of the organizational designer is to make manifest the communications links. The task of the leader is to define the direction and inspire the collectivity of people and small teams to travel to the market, technological, environmental, financial, or intellectual place the organization is destined to reach.

More has been written on leadership than on most things. The task of the leader in a telecommunications-based service provider is focused upon providing the following.

1. The compass;
2. The energy source;
3. The links between its groups.

The setting of direction is covered in Chapter 6. The notion of an *energy source* derives partly from resource allocation and partly from motivation in a wide sense. Issues of leadership arise all through the book.

If the enterprise is "well-led," then not only will the enterprise as a whole function more successfully but the individuals and small groups will work better, too. In focusing on the big picture of the organizational design, one proceeds from the principle, or assumption, that individual people and small groups are already quite good at working together. That assumption may be unrealistic and there will probably be work in parallel to be done with those elements. This is covered in Chapter 12.

5.6.2 Process-Based Organization for a T-BSP

These are considered to be the key process areas of a telecommunications-based service provider. They will have a different weight and relative importance, according to circumstances, but should provide a complete set from which to work. Each process area is described in detail in the chapter shown, with the emphasis on the necessary distinctions required or available in telecommunications-based service provision.

Two recommended forms of organization chart are shown in Chapter 12.

An overall diagram to show accountabilities and how process areas linked together are given below in Table 5.2. and Figure 5.7 An overall process map with links to the information systems is shown in Chapter 13 (Figure 13.5).

Table 5.2
Process List of the Telecommunications-Based Service Provider

Process Area	Principal Owner	Chapter
Direction and strategy/ monitor performance	CEO Financial controller	6
Marketing/intermarketing (+ stakeholder communication)	VP, marketing	7
Create and manage product	VP, products	8
Obtain order	VP, sales manager	9
Implement order	Customer project manager	10
Customer support	Customer support manager	11
Support/ and develop the organization	Financial controller CEO	12
Manage information systems	IS manager	13

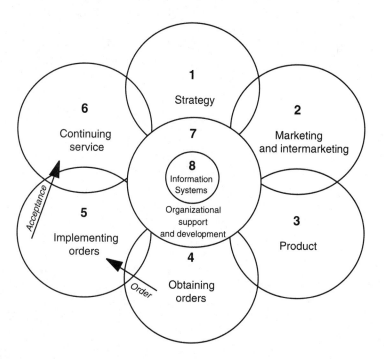

Figure 5.7 Process area map for a telecommunications-based service provider.

5.7 CONCLUSIONS

The design of the organization is a crucial factor in the success of a modern information-based/telecommunications-based organization.

No single diagram or table can describe a telecommunications-based service provider enterprise on its own. The use of several in combination is recommended.

The proposed "ideal" solution, previewed above and to be described in the next eight chapters will never be right for all circumstances. However, if the design is made around process areas, underpinned with process descriptions, tuned to reflect socio-technical issues and designed to give empowerment, then it should meet the main design needs, which are restated below:

- Strong sense of direction to guide it through change;
- Sound competitive proposition that is well understood;
- Speed in taking products to market;
- Skilled, committed people—empowered to work for their customers;
- Systematic but flexible processes and tools that save time;
- Superior teamwork. Clear roles and responsibilities.

It is to the detailed examination of each of these processes that we now turn.

Strategy and Monitoring 6

6.1 STRATEGY: PROCESS AREA DESCRIPTION

There are many books on business strategy; there is a wide range of tools. What follows is a condensed and practical set of insights, tips, and tools relevant to the telecommunications-based service provider. The chapter is designed particularly to provide the following help:

1. Focusing corporate strategy specialists on what is relevant to telecommunications-based service provision;
2. Educating people of other disciplines on the concepts and practice of strategic analysis, choice, implementation, and monitoring;
3. Providing a practical kit of tools and vocabulary for a telecommunications-based service provider management team.

The process area described in Chapter 6 covers the processes, activities, and systems that are used to select, plan, and monitor strategy. The strategy process area is the start point, and it fits into the whole as shown in Figure 6.1.

The telecommunications-based age is fast-moving, changeable, and highly competitive. Telecommunications-based service providers require a sound competitive proposition, a strong sense of direction, and the capability to respond fast to opportunity or change. Each business must understand *what precisely* it is, *why* people buy what it is selling, and why they will go on buying.

The most powerful strategies are generally simple to understand and to implement correctly. A strategy is without any value unless it is implemented. It won't be implemented if the mission does not provide a clear motivation to act. Strategic analysis comprises the matching of environment, capability, and expectations so that choices can be made.

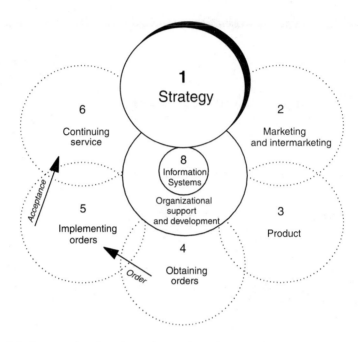

Figure 6.1 Process area 1—select, plan, and monitor strategy.

Detailed plans must be produced that will show in detail how these choices can be put into effect. These plans will cover areas such as markets, products, people, technology, information systems, and other resources. Plans must be cash- and time-bounded and they must be realistic.

Measurement methods must be put into place. After implementation, the business must rigorously monitor results and continually question the continuing competitiveness of what has been deployed. If resources could be deployed more effectively elsewhere, then the strategy is no longer a competitive one.

Often there will be no time to plan at length, for the opportunity to act may be of short duration. Section 6.8 shows how to take shortcuts in the formal processes. Section 6.6 describes the technique of maneuvering—opportunistic and instantaneous implementation of a powerful idea—as contrasted with the disciplines of the full strategic development process. Strategic planning, like many processes, works best when it is carried out by a multidisciplinary team of thinkers and doers, specialists and generalists, outsiders and insiders.

The creation and maintenance of the strategic plan is a primary responsibility of the chief executive officer (CEO). Measurement is a primary responsibility of the chief finance officer (CFO). Telecommunica-tions-based service provision is generally a fast-moving business. Situations change; the past and present are not always the best guides to the future. Multiple perspectives will be needed and information must come from a range of sources, not just one. It is vital that the planning task is not devolved away from the people and groups who really know what is happening on the ground today, but at the same time it must be recognized that the practical knowledge of these people is the very basis of bounded thinking. Focus on short-term success is vital but at the same time outside thinking is needed to consider what lies beyond the current paradigm.

Many strategies depend upon alliances for their successful execution. It is also important to take advantage of *knowledge alliances* at the earlier stage of developing strategy. The simplest form of knowledge alliance is the *network* of friends, colleagues, customers/suppliers, professional advisers, and specialist researchers. Knowledge inputs should be taken from outside the immediate technological or business focus. Chapter 1 detailed the effects of convergence of technologies, businesses, and nations. In compiling plans, senior executives should also look beyond their geographical boundaries to see opportunities, competition, and alternative methods. Consider the cases of Service providers X and Y, described below.

Service provider X, in the United States, based its business proposition on buying bulk international capacity at a discount from a range of telecommunications network operators and reselling this capacity to small- and medium-size enterprises by use of a special access number and PIN. With the continued reduction of tariffs brought about by competition, including the competition, the competitive advantage brought by the service provider was reduced. Eventually, it was below the point where it was worth the bother of both having to remember to dial the number and of paying two bills. The customers moved back to the dominant telco.

Service provider Y, with the same initial competitive proposition, looked at what had happened in the United States and decided to move up the value chain with its customers. Not only did the company continue the pressure on tariffs, but it also offered *outsourced telecommunications management* to the customers it had gained. Service provider Y managed customers' PABXs, provided bandwidth on demand, and linked in mobile, data, and directory/information services. The tariff savings are still there; but the customers are even more delighted with the savings of time and improvements in technology. Next service provider Y is going to partner with an IT company to manage customers' PCs, LAN, desktop software, e-mail connection to the Internet, and so on.

Table 6.1
Process Area Outline—Select, Plan, and Monitor Strategy

Field	Attribute
Process area description	Create strategy—analysis/choice/plans Monitor strategy
Process owner	CEO CFO
Other actors	Most of the organization, partners, and stakeholders
Processes and activities	Analyze strategic situation Identify and select from choices Set overall objectives Create and publish detailed plans Monitor results and feed back into strategy
Time to complete	1 to 3 months (then in continuation)
Inputs (and sources)	Current performance information and trend data General research information and dialogue Stakeholder views and objectives Bottom up/top down forecasts
Outputs (and destinations)	Detailed strategic plans that act upon the other process areas: providing them with environmental information, directional guidance, specific objectives, resource allocation
Support documents	*Tools*: PEST, A→B→C diagram, SWOT *Plans*: market summary plan, product/technology plan pricing/channels/sales plans, organization/people plan customer service plan, finance plan *Reports*: Finance, market share, process area performance, performance against budget forecasts, etc.
Information system lead user for:...	Strategic analysis resources Management accounting systems/monthly reports
Effectiveness measurement (measured by those receiving the outputs)	Strategy: timeliness, intellectual integrity, vision, realism, detail, acceptability. (Acid tests: Do proposed strategies 1. protect and enhance market share, 2. improve cost-effectiveness, 3. Grow new revenue streams?) Management accounting/monthly reporting: timeliness, clarity, relevance, usefulness as basis for strategy tuning

Field	*Attribute*
Risk and cost management/disaster recovery	Risk contained by obtaining as much information and deploying as much knowledge as possible from inside organization, from outside sources, and from alliance partners. Important to maintain confidentiality until ready to publish. Use people of proven commitment to the organization
	Costs (and opportunity costs) minimized by carrying out the planning and reporting process as quickly and simply as possible. Avoid manpower-hungry processes and excessive numbers of meetings. Deploy knowledge strategy (see Chapter 14) to reduce information gathering cost and to increase acceptability
	Disaster recovery: Watch for "step changes" in the environment that could wipe out competitive proposition. Rely more upon simple reports and direction statements

6.2 ANALYZE STRATEGIC SITUATION

6.2.1 The Tools of Analysis

Analysis of the *environment* is one half and analysis of *capability* is the other. There are many tools for analysis but the four most practical for a telecommunications-based service provider are probably:

1. Focused political/economic/sociological/technical (PEST) analysis;
2. A→B→C: scenario step diagram (strategic staircase);
3. Strengths/weaknesses/opportunities/threats (SWOT);
4. Analysis of comparative characteristics.

Telecommunications-based service providers do not always have a lot of time for analysis. These four tools, described in detail, are *plug and play*. They are simple to master and should save time.

But beware...although one of the purposes of analysis is to reduce uncertainty, it cannot be guaranteed...

One Future...or Many?

The wish to reduce uncertainty can result in a tendency to converge on the single most likely scenario of the future. However, many of the great events and outcomes were not predicted at all:

- The fall of the Berlin Wall in late 1989.
- IBM is said not to have seen much of a future for the PC.
- Telephones? "One day every town may have one."

6.2.2　PEST

The PEST considers political, economic, sociological, and technical factors in the environment. A PEST analysis may be based in the present or in the future. Each of us has a perspective on the PEST, and there is no single future. Chapter 1 serves as a PEST for the business of the telecommunications-based service provider in general. But, whilst analysis of that scope can grab interest and create new ways of looking at things, it is essential that the telecommunications-based service provider focuses in on a specific environment or situation where a profitable business can be grounded.

Table 6.2
General and Focused PEST

General PEST—Next Few Years	*Focused PEST—German Telecommunications*
Political	*Political*
Deregulation and privatization	Deregulation of telecommunications - 1998
Trade blocs: Americas/Europe/Pacific Rim	Privatization of Deutsche Telekom 1996
Stable governments, increased democracy	Continued harmonization of European
Harmonization of laws and practices	legislation
	Ever-stricter environmental protection laws
Economic	*Economic*
Gradual increase of prosperity	Unemployment rising, but wages rising too
Rising inflation	Continued differential development
Unemployment rising in most regions	between former East/West Germanys
Disposable income rising	Deutschmark weakening against $ and £
	Anti-inflationary policy

General PEST—Next Few Years	Focused PEST—German Telecommunications
Sociological	*Sociological*
Population growing but also aging	Reduced wish to work long hours, but
Increasing mobility between countries,	continuing pressure to do so
social classes	Skill shortages. Greater tendency to
Growing consumerism	outsource to specialists
Passenger traffic increasing	Increased spending on services and leisure
More time spent watching television	Increased propensity for physical sport and
	games of all sorts
Technical	*Technical*
Increased role of telecommunications	Telecommunications tending to very low
Rates of obsolescence increasing	cost
Machinery becoming ever cheaper	Computing tending to unlimited power and
Tendency, nonetheless, for more and more	storage capability
complexity to be built in to devices	Devices becoming increasingly easy to use
	and widely available

Table 6.2 shows clearly that the more focused the analysis, the more useful it is for a single company strategy. But a telecommunications-based service provider should never just consider its immediate environment but also those of its suppliers, customers, and analogous areas too. Take the example of telecommunications-based service providers X and Y in Section 6.1.

The main strategic issues identified by telecommunications companies in general are: competition, deregulation, falling costs and prices, downsizing, adding value, globalization, market focus, strategic alliances, and what to do about the Internet.

The main issues facing small- and medium-sized enterprises are fierce competition leading to a greater and greater focus on core propositions and a need to extract maximum value from every asset. But, whilst this has caused some focus on cost reduction, the increased market opportunities and skill shortages are making it more difficult to keep staff skills up to date and are increasing the propensity to outsource noncore activities (such as telecommunications management).

Computer manufacturers and maintainers, systems houses, and value-added resellers see decreasing margins on equipment and want to provide more services to their customers.

T-BSP X didn't look beyond the paradigm of the present. T-BSP not only considered the future from the standpoint of its main supplier and its typical customer, but it also looked at the impact of a changing PEST for a parallel type of business and spotted an opportunity to strengthen its competitive proposition through partnership.

Environmental analysis and its applicability to the development of niche opportunities in telecommunications-based service provision is considered in more detail in Chapter 8.

However, in the telecommunications-based age there is a powerful tool for market research: the Internet, and especially the home pages it contains. As a minimum, the strategist should be downloading the home pages of competitors. Discussion of *futures* and insights into the issues interesting to potential customers are also there for study and incorporation into strategic thinking.

In Chapter 7, the use of the Internet as a marketing and stakeholder communication tool is considered in a little more detail. However, the strategist who wants to learn how to get on to the net and start looking for information can refer to many books and sign up with an Internet service provider. There can be very few specialist researchers or librarians who are not now using the Internet as a primary information search resource.

6.2.3 A→B→C Scenario Step Diagram

A PEST may be derived from discussions with a wide group of people inside and particularly outside the immediate business. It may be written up as an essay, as a scenario (for example, see Chapter 1), or as a series of directional statements as in Figure 6.2. Some groups like to create visual representations of the future compared with the present. These mechanisms work well as an approach to the issues in question. They ensure that *overall directions and influences* are identified as well as possible. They assist in plotting the *edges* of a specific strategic environment. But for the next focusing stage, a more focused tool is also needed.

It may be easier to assess the impact and identify an opportunity space if it is expressed in an A→B→C scenario diagram, or "strategic staircase." A is used to describe the most relevant aspects of the present scenario. C portrays a future point (typically two or three years away). B shows the situation in 6 to 18 months time. The telecommunications-based service provider assesses the changes taking place and positions product creation or development accordingly. Alternatively, use last year/today/next year/3 years from now, as shown in Figure 6.2.

6.2.4 SWOT

The SWOT analysis is a tool that maps the characteristics of the strategic environment against the characteristics of the firm. A SWOT was developed in Chapter 3 that identified the strengths, weaknesses, opportuni-

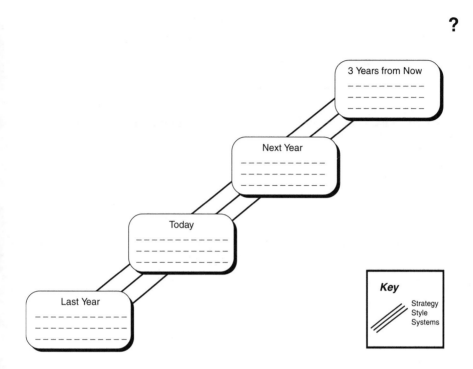

Figure 6.2 Strategic staircase (A→B→C scenario diagram).

ties, and threats facing mobile telecommunications service providers in the United Kingdom. For convenience it is repeated here (Table 6.3).

If you can consider the different PESTs seen by suppliers, yourself, and your customer, you can also consider different SWOTs. The simple yet powerful insight is that the weaknesses of other members of your extended value chain provide and opportunities for you to provide a solution. (See also Section 9.8.4.)

SWOTs are quick and easy to do. Every one in the telecommunications-based service provider's organization should be familiar with the main points. The SWOT is one of the main guides to *what* the telecommunications-based service provider should be doing.

6.2.5 Analysis of Key Characteristics

The key characteristics analysis will help decide *how* the telecommunications-based service provider should be taking its products and services to market.

Table 6.3
UK Mobile SPs: Strengths/Weaknesses/Opportunities/Threats

Strengths	*Opportunities*
Customer base (and database)	Continued growth of the market
Going concern: Premises, trained and experienced staff, systems and processes	Sell out to network operator
	Work with partner who can extend value
Generally entrepreneurial attitude	proposition (N. B. with a fixed network
Brand (possibly)	operator)
Familiarity with mobile communications	Rebrand in association with a partner (e.g.,
Existing partnership relationships	blue chip X service provider)
	Find new products for same market (e.g.,
	data, value added services)
	Find new market for same product
	(overseas)
	Find characteristics of best 20% of
	customers; create heavily enhanced value
	proposition for them at a premium price.
	Dispose of other customers and scale down
	operation accordingly
	Do new analysis of strengths and redeploy
	into another form—e.g. a "super dealer"

Weaknesses	*Threats*
End of protected position	Decreasing margins
Lack of cash	Hostile take-over
Unprofitable	Increased competition from other suppliers
Weak negotiating position	of core product
Lack of system/solution kills.	Inaction during time remaining
Underinvestment in systems	
Brand (possibly)	
Insufficient scale to compete with network operators	

Further focusing reduces the risk of setting out on a path where the odds are stacked unfavorably or of being "outsmarted" by the competition. The SWOT should be considered in association with an audit of key characteristics of the firm, its competitors/partners, suppliers, and target customers. Matches and differences will give important clues when deciding the best method of addressing a business opportunity. The listing should cover such areas as the following:

- Product;
- Size;
- Market share;
- Geographic coverage;
- Hours of operation;
- Technology platform;
- Distribution channels;
- Branding;
- Pricing strategies;
- Knowledge base;
- Cash resources;
- Skills;
- Major challenges.

If the listing covers not only the potential players but is also used to characterize potential opportunities, then shortfalls or overcapability will quickly be identified. Such an analysis then gives a powerful impetus to working with partners. Consider the following case.

A systems house that has a deep technical knowledge of data communications, wide area telephony, and database management systems plus a respected presence in the government and large corporate market decided to bid for a contract to design, build, and operate a national lottery. The corporation assessed that this largely telecommunications-based opportunity was within its capability. Rigorous analysis of the respective characteristics of firm and opportunity showed a shortfall of skills and experience in mathematics, lottery promotion, and retail channels to market. Partners were selected to cover these areas and the resulting consortium launched on time and within budget. In operation, it went from strength to strength.

6.3 IDENTIFY AND SELECT FROM CHOICES

Having constructed an analysis of the environment using some or all of the above tools, the telecommunications-based service provider must identify and select the best basis on which to convert analysis into reality and the best ways to manage the risks coming from the competition. Description and quantitative analysis of a value chain for your own organization and for your competition will soon, also, identify bases of advantage or disadvantage, in much the same way as the analysis of key characteristics covered at the end of Section 6.3.

Traditionally, there have been two main groups of strategies for an enterprise: They are cost focus and differentiation focus.

Cost focus strategies depend upon prices to be set at levels that cannot be achieved by the competition. In practice, prices would have to be set at least 15% apart to make a perceptible difference to market share. Strategic breakthroughs in bringing down the cost of significant items in the value chain will cause a major shift.

The introduction of direct telephone-based insurance as an alternative to local branch offices caused a fundamental shift in the prices that could be charged in the market as a whole.

It would be difficult for a telecommunications-based service provider to compete with a telco in constructing a trunk network; a manufacturing operation in Taiwan or Lithuania is likely to have lower labor costs than one in Paris or Los Angeles.

Sales channel costs of an airline can be reduced considerably by setting up a worldwide back office for its booking service in a low-cost region, such as the Philippines or Caribbean.

Differentiation strategies are those that entail a perceived added value in the eyes of the customer. This perception will either provide a market share protection mechanism or else an opportunity to charge a premium price.

GSM mobile telephony networks charge more than personal communications systems (PCSs) for phone calls. Why? It is because they were positioned in the customers' minds as business-oriented services; the GSM is said to be better at hand-off from one cell to another; GSM provides international roaming.

It would be difficult for a telco to achieve a better understanding of the financial information marketplace than had already been gained by the highly focused Reuters or by Standard and Poors.

A style-branded product or service commands a premium over a nonstyle-branded product.

Customers will pay more for what they perceive as "better service."

In the fast-moving world of telecommunications-based services, a third generic strategy is also available: *first-to-market.* Customers will not normally move for price reasons unless the differential is over 15%. A large amount of research has been devoted to showing that customers tend to leave their suppliers because they do not feel valued. Thus if a telecommunications-based service provider can get to market first and also project a good service image, then it follows that a price premium of up to 15% is available. Some observed attitudes follow.

- "I've been with them from the start."
- "I'm used to them."
- "They haven't got all of the features of product B, but I can't be bothered to learn all that new stuff."

Another way of looking at the same thing is to say that customers regard what they buy as either a commodity or else as a value added product. Rational behavior dictates that customers will ultimately buy what they perceive to be the cheapest or most convenient commodity that meets their requirement. On the other hand, the buying decision for a value-added product is based upon a much more complex set of values. Some are clear: supplier has a better understanding or a more comprehensively functioned offering. Others are much more subtle—as discussed in Chapter 7—and at the bottom line for the telecommunications-based service provider is always going to be included the "service" given.

What is disastrous is to set out on a strategic path where the product is both *un*differentiated and at a *higher* cost than the opposition.

6.4 SET OVERALL OBJECTIVES

The overall objectives of an organization can be expressed in many different ways. Classic organization theory says that a commercial organization should act to maximize the value of its ordinary stock, which should equate to the net present value of its future earnings plus the disposal proceeds of its assets.

It is not always easy for people in the organization to know what is happening to the stock price—though more and more managers are given precisely this measure as an incentive, both short term and longer term. In practice, organizations tend to act to provide a balance of benefits as well as money for their shareholders: fund providers, employees, and customers.

Strategic logic says that a business must protect and enhance market share, reduce costs, and develop new streams of revenue. This is helpful but, again, too complicated for most employees to understand. Organizations often take weeks of thought and then produce a paragraph or even more of high-sounding aims and values. They don't help much, in my opinion.

What is needed is an approach that energizes employees to act in a focused, united, purposeful manner. The good leader sets a mission and paints a picture of where the organization is going and what things will be like when it arrives at its goals. Also, people like to know what is expected of them; they like to have a stake and influence in what is going on and they insist on clear roles and responsibilities. Most of all, people want to know what the objective is, what it is that they can follow whilst knowing that colleagues are following it also.

The best objectives are those that are precise, realistic, measurable, time-bound, motivational, easily understood, and actually remembered. It has been said that a good objective is also one that is short enough to be printed on the front of a tee-shirt. Here are some examples of good directional statements:

- "We will put a man on the moon by the end of the decade, and bring him back alive."
- "Get more calls answered better, faster."
- The service provider might want employees to know that "nobody will go far wrong if they cause customers and potential customers to increase their level of dealing with us."
- Pay off what we owe by the end of the year.
- Put more calls on the network.
- Give the production department a better telesales service than they can get from an outside contractor.

More directly, the aim can be expressed simply in terms of beating the competition:

- Earn $100m pa revenue by 5 years from now;
- Win the business of five of the top eight banks.

6.5 CREATE AND PUBLISH DETAILED PLANS

However, these inspirational objectives are of no use if they are not underpinned by credible, consistent, and detailed plans. There is little that is different in a telecommunications-based service provider from any other organization in this regard.

What does distinguish the world of the telecommunications-based service providers is that they may not get the plans written down tidily before the end of the year to which the plan relates. Notwithstanding, all of the issues must be understood and addressed in some way or another, however sketchily.

The set of plans should, almost certainly cover the following areas:

- Overall business objectives—PEST, SWOT, analysis of key characteristics, market trends, partnership possibilities, basis of generic strategy to provide competitive proposition (e.g., "lowest price," "most knowledgeable, and highly focused on off-road adventure trips in Canada");
- Financial plan—revenues, costs, capital, cash flow, profit, and cost center budgets;

- Marketing—market size/attractiveness analysis, target markets, market communications, brand management, market development, promotion;
- Product—price/cost, features, benefits, competitors, forecasts, new products, and enhancements;
- Sales—targets, channels, sales incentives, major bids, major customers;
- Capacity/project plans—schedules, forecasts aligned with sales plans;
- Customer service—service levels, opening hours, response times, time to answer/time to clear faults;
- Organization development—teams , people, recruitment, training;
- Quality management systems—processes, documentation, quality certification;
- Information technology—call centers, fixed and mobile telephony, automated call distribution (ACD), interactive voice response (IVR), computer telephony integration (CTI), desktop computing, work-flow management, decision support;
- Disaster plans—how will the organization respond in the event of fire, flood, computer crash, loss of key employees, major financial default by a customer, consequential loss claim, product liability claim, personal injury claim.

6.6 CONDUCT MANEUVERS

6.6.1 Precision Versus Speed

It is not always possible for telecommunications-based service providers to conduct a long planning process. An opportunity may appear for just a short while. There has to be a quick but reliable method of knowing whether the opportunity is good or bad and of then of exploiting it in the short time available. We live in an uncertain technical and marketing environment. It is changing and growing rapidly. What practical approached will help us to manage change, technical development ,and explosions of choice for our customers? How can we act more quickly? The technique to do this is called maneuvering.

Traditional planning approaches proceed through PEST and then into formal SWOT assessments for ourselves and for our competitors. Careful measurement of market size follows. A financial model is constructed. Often this plan is fed into higher and higher level corporate plans and then, once the budget round starts, all of the revenue projections and resource allocations are stretched and compromises are reached.

Normal prudence on the part of the corporate planners will ensure that the plan is intellectually sound and arithmetically proven. Unfortunately, a strictly logical intellectual approach can also yield the same conclusions for all of your competitors, so the market becomes crowded with everyone doing the same thing. Aren't most of the world's largest suppliers looking at the same, top 1,000 multinational corporations?

So, all may not be well with this approach. Meanwhile, the company's latest change initiative is bringing a new vision to the surface. By the second year the whole thing is out of date. The originators of the plan have moved on to new jobs. Besides all the board really wanted was to reduce the gearing that year...

Being "right" may be fatal. Sometimes it's better to do things that may appear to "go wrong" but are going in the right direction, make some margin, and are being done now. The problems can often be ironed out later, once some achievements have been made.

We actually don't have to have a grand plan complete in every detail over time. We don't have to stick to a plan. Plans are only self-fulfilling in a closed and centrally planned economy or under conditions of total monopoly or monopsony. In times of uncertainty and high rates of change, the comfort engendered by a plan may be illusory.

We can, instead, undertake a set of maneuvers that each provide benefit and are appropriate for the future we see and for the future that emerges and comes to us. A new approach is therefore open to us. It comes, unsurprisingly, from California's Silicon Valley.

Maneuvering works like this: First, you must have a clearly articulated vision of the world ahead and of your mission within that vision, as discussed earlier. You will, therefore, understand your competitive proposition and ways of doing things in a defined market with the knowledge and capabilities you have at your disposal.

6.6.2   A Good Maneuver

How do you do it?

1. Spot the opportunity.
2. Implement it quickly.
3. But take insurance about it going wrong.
4. Improve the solution.

What makes it good?

- Cash positive;
- Simple and probably quick;

- Moves you in an appropriate direction to a better market position;
- Lies within your area of knowledge and focus;
- Doesn't confuse anyone other than your competitors;
- Doesn't prevent you from executing a better maneuver by creating an exclusive partnership, by pre-empting resources;
- Can be executed better and faster by you than by your competitors;
- Has containable risk—a fallback solution;
- Doesn't damage your core business;
- Is for an idea whose time has come or, by maneuvering, will cause that to be.

6.6.3 Maneuvers in Practice

Clearly, maneuvering is going to be easier for a smaller organization, just like a fast-moving service provider but one of the most inspiring examples of a maneuver, discussed in the following paragraphs, comes from a large American corporation.

A well-known long distance operator decided that it was probably worth launching a new tariff to stimulate people into making even more calls to those they call most frequently—close friends and other family members. Instead of spending two years developing the product, the company simply shot an advertisement that explained the concept and showed it to the billing department. "This ad goes out nationwide on prime time TV in three months time. Now go and find a way of billing the service to our customers and maybe then to a million more."

The product was launched on time. For the first few weeks, bills were prepared on a hastily configured PC. The rest, as they say, is history.

6.7 PARTNERS

Partnership is one of the key tenets of telecommunications-based service provision. The concept is based upon taking the information and technology elements from producers and providing a channel to a selected and then nurtured customer base.

The role of a partner is to provide or enhance aspects of the value chain that cannot be provided better from within the organization. Thus, partners may provide the following.

- Technology;
- Products;

- Information/knowledge (including law, tax advice, etc.);
- Intellectual property;
- Brand;
- Cash/loans;
- Political influence, access to decision makers;
- Channels to market—marketing and sales;
- Extension of geography—by being based in another place;
- Distribution and support—warehousing, transport, installation, maintenance and repair;
- Extension of hours—either by being based in another time zone or else by providing *extended hours coverage* by deployment of existing arrangements on a marginal cost basis.

There is a fine line between a supplier and a partner. The terms under which a *partnership* is conducted may be one of the following:

1. Simple supply of information, goods, services or perhaps licenses for cash.
2. Enhanced supply arrangement, whereby the partner provides added value, such as financing or additional technical advice/sales support. It is quite likely that this partner will share in the fortunes of the business—bearing risk or obtaining reward. The basis of payment may well be that the partner takes a share of revenues as they come in.
3. A formal distributor agreement, or something similar.
4. A formal joint venture arrangement.
5. Full equity partnership.

It is normally better to explore these possibilities in ascending order. With each step up, the arrangement shares more risk and reward. But with each step up there is also more responsibility and the partnership becomes more complex to operate. It does not necessarily follow that commitment or focus is always increased, too, though there is clearly an increased degree of exclusivity and reduced competition.

In every case, the following issues must be rigorously discussed and documented:

- What is the partnership seeking to achieve in the marketplace?
- What are the expectations in terms of volumes, revenue, competitive proposition, profit, and so forth?

- Why are the partners working together? How did they choose each other?
- What are the respective roles and responsibilities?
- How will the relationship be managed and by who?
- How, when, and under what circumstances will the partnership be dissolved?

6.8 LEGAL AND REGULATORY ASPECTS

It is well beyond the scope of this book to give legal advice. However, the workings of normal commerce, sale, contract, payment, and so forth should be reviewed thoroughly, especially before setting up business transactions over new media such as the Internet.

What law will actually apply if the business trades across national borders? Will the licenses embedded in the telecommunications-based service providers product be valid for a wider market?

Are their any special considerations regarding content, especially those of such factors as cultural acceptability and decency?

These aspects must be thoroughly checked out before committing energy and resources to any project. The law is developing very rapidly in all of these areas. Not surprisingly, there is a growing number of lawyers who specialize in the area of electronic trading and similar issues. It may be appropriate to use a law firm with an established *international* reputation and presence.

It is also impossible to give advice on the regulation and licensing of telecommunications-based businesses for each of the markets that may be encountered. However, telecommunications-based service provision is normally based upon the tenets of competition and of deregulated access to customers on equal terms with former monopoly suppliers. It is imperative that the real market situation be properly understood.

At the very least, consider the risks inherent in taking customers from a telco that may in turn be supplying a vital part of the connectivity and transport of the product in question. Former monopolies can cling tenaciously to what they consider to be their rights; they may not treat telecommunications-based service providers in a totally even-handed way. Regulatory experts are not always lawyers; it is commercial thinking skills that are needed above all else. It may be helpful to seek advice from an expert in a country that has gone further down the road to deregulation, since more of the pitfalls and opportunities will have been experienced already (e.g., Australia, the United Kingdom, or the United States may have more experts than other countries in their respective regions).

6.9   MONITORING RESULTS

All businesses have to monitor results and review progress towards strategic objectives. This function is no less necessary for telecommunications-based service providers.

Classic monitoring methods include the following:

- Setting objectives for each team and checking progress monthly via a written report;
- Board meetings as milestones;
- Management accounts.

On objectives, there is little to say that is particularly relevant to telecommunications-based service provision and that alone. It is certainly important to recognize the fast-moving nature of the business and the fact that people at the customer-facing edge of the business are the best judges of performance elsewhere. Committed managers will undoubtedly have going a "scoresheet" of their own. Such a scoresheet, be it financial, quantitative, or anything else, will have a high degree of validity and is probably the best measure to use! Reports, when produced, should be grouped together, interpreted, and summarized to show the whole picture and then circulated as widely as considerations of confidentiality allow.

A board meeting for a telecommunications-based service provider is unlikely to be any different from a board meeting elsewhere. Members should invariably be users of the relevant products and services. They should telephone in regularly to maintain an up-to-date measure of customer service standards.

Management accounts are frequently constructed badly. Often, budgets are not agreed till several months into each year. Whereas financial accounts focus on the corporation as a whole, management accounts should be based upon its constituent teams—*profit centers* and *cost centers*—so that the managers of these teams can review results and trends at the organizational level at which any necessary management action can be directed.

Putting each transaction in the following form will assist in creating management accounts and will also provide an audit trail for cost control purposes.

Period	"Month 4"
Cost Center	"A111: Customer Service East Coast—Field Engineers"

Description	"1234: Travel Expenses"
Value	"$400.35"
Reference	"Expense Report: J Brown"

Individual profit and cost centers will be aggregated into bigger units. Management accounts should include key operational data and data shown in dollars. Table 6.4 is an example.

Table 6.4
Typical Management Accounting Report

Management Accounts for Month Ending 29 Sept. 1997
(Financial figures in $K)
Cost Center:
A111, Customer Service East Coast—Field Engineers

Description	*This Month*			*Year to Date*			*Full Year*		
	Actual	Plan	Last Year	Actual	Plan	Last Year	Forecast Actual	Plan	Last Year
Customers This month and at year end	12,003	11,000	8,778				18,900	17,000	9,888
Equipment	102	120	60	805	800	404	1,600	1,500	900
Information Services	58	60	40	268	200	124	395	400	190
Maintenance	71	70	50	127	100	72	405	400	110
Revenue(a)	231	250	150	1,200	1,100	600	2,400	2,200	1,300
Cost of Sale(b)	111	115	67	528	540	294	1,032	1,012	631
Trading Profit (c)= (a)–(b)	120	135	83	672	560	306	1,368	1,188	669
% of Revenue (d) = (c)/(a)	51.9%	54%	55%	56%	50.9%	51%	57%	54%	51.5%
Salaries etc.	50	60	40	250	260	240	500	520	480
Travel	8	6	4	40	35	24	85	70	60
Rent	20	20	20	120	120	120	240	240	240
Other	11	10	25	45	40	50	80	68	100
Total Operating Costs (e)	89	96	89	455	455	434	905	898	880
Contribution (f) = (c)–(e)	31	39	(6)	217	105	(128)	463	290	(211)
% of Revenue (g) = (f)/(a)	13.4%	15.6%	(0.04)%	18.08%	9.5%	(21.3)%	19.29%	13.18%	(16.2)%
Head count (This month and at year end)	8	10	6				12	14	11

6.10 SHORTCUTS AND CONCLUSIONS

By way of a conclusion to this chapter on setting and monitoring strategy, the content and knowledge has been summarized as a "shortcut."

If in doubt, a *simple* strategy should be produced, following the defined processes but doing them as a set of sketches that build up into the quick overall picture:

1. Present situation, future situation.
2. What will the market want? What can we probably do better than most (e.g., being cheaper, adding more value by knowledge of a niche market)?
3. Describe how the market will perceive us, when we have succeeded.
4. Competition? SWOT? Any show-stoppers or ways we can reduce risk?
5. How will we produce, market, and support the project?
6. Illustrate the strategy with some "maneuvers" (see Section 6.9).
7. Alliances?
8. Finance and resource plan.
9. Actions?

Produce the strategy in a day or a week and not a month or three months. Using these timescales, the whole mind is focused upon the task and the whole picture is grown and processed in right brain as well as left.

Ask a few outrageous questions: "What would happen if all of the products we're producing today for a good margin were being given away in five years time?" "Imagine a new extraordinary combination of ideas such as vegetarian/sausages, tropical holidays/Central Manchester, centrally held book/printer on university campus/credit card..." "Imagine telecommunications were free; what would that do to the way we sell our product?"

State current position, the position you want to reach, some points along the way, using an ABC diagram—as in Section 6.2:

Subject strategy to the three acid tests:

1. Do the plans protect and enhance market share, improve cost efficiency, grow new streams of revenue?
2. Is the competitive proposition truly compelling? Why is it customers will buy from you rather than the competition? Why should they buy at all?
3. Is the strategy realistic? Does it have the support of those who will make it a reality?

Keep going back over the early stages. Make the strategy simpler and more comprehensible. Link it to a few big ideas. But be sure to do the sums...

By the time you come to present your strategy formally, preferably as a two-hour presentation and not as a 100-page document, you should have a very clear picture indeed of what you need to achieve and how you are going to do it. You should have, too, a clear plan for ensuring that everyone concerned ends up with a common vision of what is to be done.

Keep people up-to-date with progress through the year, with progress towards stated objectives.

Dialogue with customers, suppliers, potential partners, and specialist advisers will enhance strategic thought from the outset. Once drafted, the strategy must be discussed and agreed. Finally, a strategy to sell a service will come to nought if the product in question does not register with the market to which it directed. These issues are the subject of the next chapter.

Marketing/Intermarketing

7

7.1 SCOPE

In the telecommunications-based age there are many powerful marketing media available to everyone. *Intermarketing* is a term to distinguish what can be done over the Internet from what is possible by more classical means. The purpose of this chapter is to describe the vital role of marketing and intermarketing for the telecommunications-based service provider, to list the key processes, and then to briefly describe those that are supported particularly well by new media methods.

The first purpose of marketing and intermarketing is to attract attention to the competitive proposition and then sustain dialogue around it.

Marketing/intermarketing exchanges information between most of the other process areas and the marketplace. The marketing/intermarketing process area particularly follows *strategy* and precedes *product* and *orders,* as shown in Figure 7.1.

The telecommunications-based service provider must motivate people to buy its products and services for the first time. It used to be said that "if you built a better mousetrap, then the world would beat a path to your door." It is no longer the case. With so much information, with so much choice, and with such rapid change, it has become extremely difficult to gain the attention of potential buyers. It is equally challenging to convert that attention into a worthwhile prospect of obtaining an order. The processes to do this are described—especially in the context of the telephone and the WWW.

For a service provider, however, branding is not just about an attractive Web site, glossy brochures, and memorable prose. There's another factor—"moments of truth." Everyone contributes to marketing. Telecommunications-based service provision is, generally, a transaction-intensive operation. This means there are many dialogues taking place in the course of a single day, especially between customers and customer-facing

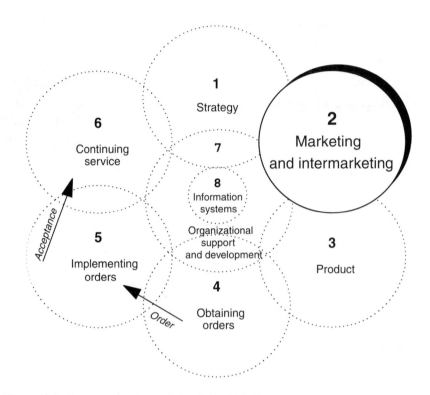

Figure 7.1 Process area 2—marketing/intermarketing.

people. These contacts have been called "moments of truth." If this is where communication with the outside world is largely occurring, then it follows that the *brand* and its values are in the hands of *everyone*, rather than just those of, say, the firm's advertising agency or simply its marketing department. With a telecommunications-based service proposition, the customers will probably face increasingly automated contact such as via a network interface, call-handling equipment, or even through the basic communications billing.

So, it is with both potential new buyers *and* with existing stakeholders that the service provider needs to develop and foster a common, favorable view of the product, the organization, and the future. That is the second role that marketing/intermarketing has to fulfill in telecommunications-based service provision.

There is a real communication problem to be solved for each participant in maintaining congruence of expectations and reality and also to explore new possibilities together. Stakeholders, be they stockholders, employees, existing customers, or suppliers, are all routinely subjected to excessive information. With *information* increasingly seen as the fourth factor of production, dialogue amongst strategic partners is an important ingredient in developing and maintaining the telecommunications-based service provider's stock of knowledge. Each of the eight process areas contain tasks designed to accomplish a part of this. The role of marketing/intermarketing is to focus these for target audiences and to draw the individual communications together with a single timetable, formatting, and message.

Finally, the marketing/intermarketing team is the natural place to maintain the *library* of information on the market and on competitors for the people and teams in each of the other process areas.

7.2 PROCESS DESCRIPTION

Table 7.1 describes the process of marketing/intermarketing processes that have to be carried out to support the overall objectives of the telecommunications-based service provider.

Table 7.1
Process Outline Table—Marketing/Intermarketing

Field	Attribute
Process area description	Marketing and intermarketing Stakeholder communication
Process owner	VP Marketing CEO
Processes and activities	Positioning the product and the brand Generating sales leads and supporting sales channels Measuring perceptions; managing stakeholder strategic dialogue Central information function (part of)
Time to complete	Communication should be continuous and dialogue continuously updated. Marketing and intermarketing will include many individual "campaigns" of short or medium duration.
Inputs (and sources)	Strategic objectives and plans (strategy) First cut product plans (product) First cut sales plans (sales) Formal market research, soft market research, general perceptions, etc. Media inquiries General inquiries, product inquiries (external and internal)

Table 7.1 (Continued)

Field	Attribute
Outputs (and destinations)	Marketing plans—including special studies (various) Marketing material of all kinds to stimulate inquiries about products and services and support sales (obtain orders) Reports on brand strength, customer attitudes, etc. (strategy) Press releases Corporate identity standards (support/develop organization, manage information systems)
Lead user for these systems...	Marketing communication including intermarketing, Web site authoring, etc. Sales lead generation Marketing collateral—brochures, etc. Market research resources
Support documents	Corporate Web site Market research and competitor information Marketing material: from each of main lead generation mechanisms Corporate documents, such as annual reports, press releases
Effectiveness measurement (measured by those receiving the outputs)	Sales leads passed to Sales Positive shifts in perceptions of audience Also: consistency in direction, content, production standard, and layout of all public material
Risk and cost management/ disaster recovery	Risk: ensure that all concerned are briefed on overall directions, key issues, and areas of sensitivity. Use approved messages and text. Risk: use trained communicators. Ensure there is no incorrect use of trade marks

7.3 THE CENTRAL ROLE OF THE WEB SITE

A Web site can be used for any or all of the following:

- Advertising the firm and its products;
- Handling general inquiries about them;
- Providing information on prices, delivery, local support arrangements;
- Order input;
- Delivering the product (e.g., as a download of free software);
- Customer surveys;
- Public relations statements.

Each of those areas will be described as it is reached in this and succeeding chapters. Further information will be found in the many new books that are now appearing devoted to the subject of intermarketing.

7.4 POSITIONING THE PRODUCT AND THE BRAND

7.4.1 Brands

Part of the strategy process area covered the assessment of the organization's capability. Part of the capability (covered in the SWOT) will depend upon how the firm is perceived in the eyes of the world. The *brand* is like a "perceived personality." The brand is a general term covering two things that customers and others experience: *the value proposition of the product* and *the perceived behavior of the firm.*

In the telecommunications-based age of instant and widespread communication, the competitive edge will come from the communication and from the value proposition created and maintained, perhaps even rather more than from the inherent qualities of the product itself. A familiar product is probably seen as a safe purchase option. But for a new branded proposition to achieve easy and comfortable recognition in the customer's mind is a significant challenge.

7.4.2 Being Memorable

If asked to name the top ten car rental organizations, soft drinks, or mobile telephone manufacturers in the world and to say something significant about them, most people would be hard-pressed after passing the number three contender.

Names and descriptions are easier to recall if there are various existing clues; for instance, if they are generally classified by a state or country name, they will already have significance and structure in the memory. So, if somebody is asked to name ten airlines, newspapers, or phone companies and say something about them, then most of us would complete the task with ease (American Airlines, British Airways, Air New Zealand, Finnair, New York Herald Tribune, Straits Times, Frankfurter Allgemeine Zeitung, Bell Atlantic, Hong Kong Telecom, France Telecom).

However, although many airlines, newspapers, and phone companies proclaim their brand by geographical or linguistic association, not all of them do. *Value*Jet is saying something about its price proposition; *Virgin* is building upon an already established lifestyle brand; *Sprint* and *Mercury* may have been emphasizing *fast* technical advantages. Some organizations proclaim a "world" message that emphasizes the spread of their market. The names of the loyalty clubs of both airlines and telcos often are designed to appeal to vanity and desire by recipients for a measure of special treatment. Newspapers and journals, of course, are even more commonly named by their subject.

In the telecommunications-based era, the name or message has to be strong, simple, and memorable—a sound bite is easier to remember than a paragraph. Consider the following sound bites:

- "Just do it," "The real thing," "Fly the Friendly Skies," "Singapore Girl."
- Phone companies use catch phrases to encourage people to make more phone calls: "Be there at your child's bedtime," "It's good to talk," "Call the folks back home."
- "We search the net for business information so you don't have to."
- "First Virtual" (bank).
- "Bookshop."
- "The Internet School."
- "First Call" (theater bookings).
- "Drinks" (mail order liquor store).

7.4.3 Being Accessible

If you are going to advertise, you must be seen where your potential customers will be looking.

Saluting the telecommunications-based age, some airlines are now painting their telephone number all over their planes—the rationale is that it is the only thing that has to be known by the customer to connect with the product.

More and more firms, too, are advertising their WWW site locator and e-mail address. This promotes an image of openness, accessibility, and being truly up-to-date.

This strategy can be extended if the Web page is a virtual storefront set in a virtual shopping mall. If the product is sail training course, then create a hypertext link with the home page for a boatyard or holiday resort. Make sure the home page is registered on the principal search engines. Advertise the home page on billboards—both within the WWW and at the end of the street.

Intermarketing is global. Customers may be located anywhere. The "drinks" liquor store found a previously unconsidered market of its own nationals living and working overseas. Wider markets may or may not be advantageous, depending upon support arrangements, pricing strategies, license arrangements, and many other factors. However, before closing off possibilities, it pays to remember that new competitors now have access to home markets, too.

Traditional methods of advertising are shallow in content, one way, and broadcast to a relatively narrow market. Even prime-time television advertising washes over most of the people that it touches. Intermarketing, however, can be as detailed as you like—peeling back more and more layers of the onion from the original attention-getting device. Intermarketing is interactive, so that each recipient can decide *when* to engage and can decide *how* to engage, selecting the information that is relevant to them.

Few marketing communications strategies will be now considered complete if they fail to include a World Wide Web site that will enable interested parties to learn more about a firm and its products. The use of color, compelling graphics, hypertext links, and interactive response mechanisms (i.e., a PC or network-computer based system identical in concept to interactive voice response and database inquiry) will provide Internet callers with what they require. However, it must always be remembered that there is a time penalty for each added feature or graphic.

A telecommunications-based service provider can take one of several approaches in creating a Web page. For example:

- Do it yourself using an appropriate PC-based software package or professionally developed software associated with larger systems;
- Use a specialist Web site author;
- Go to its Internet access service provider and ask for help.

There are now specialist advertising agencies to advise on all of these issues. Find them on the Web.

7.5 GENERATING SALES LEADS AND SUPPORTING SALES CHANNELS

Most lead-generation activity can be defined as a *campaign;* that is, a series of linked marketing activities directed at generating leads. With intermarketing, there are other special considerations, too.

Marketing campaigns will typically address thousands of possible customers. The campaigns are designed to trigger action. Sales generation is both marketing and sales. A useful dividing line can be drawn; for example, when a prospect is qualified to the point where there is more than a 10 to 15% chance that a sale will result within, say, six to nine months. After that, the prospect should be held on a sales tracking system and nurtured to success. The sales generation process can be illustrated as a funnel, as shown in Figure 7.2.

Here are examples of campaign mechanisms:

- Advertising a specific selling proposition;
- Special offers;
- Conferences, shows, and seminars;
- Direct mail;
- Direct response advertising;
- Database marketing;
- Telemarketing;
- WWW based self-selection.

All campaigns should be thoroughly planned in advance. The *tactical brief* is a tool that will focus effort and save money. It comprises the following:

1. Title;
2. Target market (size, structure, and characteristics);

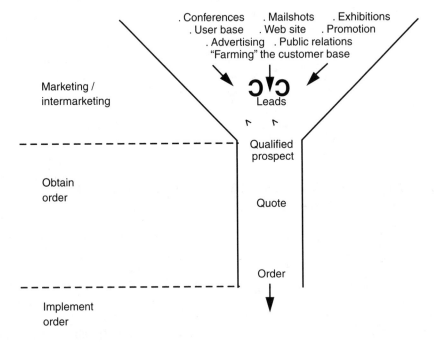

Figure 7.2 Sales generation funnel.

3. Overall objective (quantified and time bound);
4. Messages;
5. Budget and timing;
6. Description of alternative mechanisms considered;
7. Measurement—"how will we know if we've been successful?";
8. "Take out" (what the target will be thinking afterwards);
9. Action desired of target;
10. Resource plan, accountabilities, and so forth.

All of the listed mechanisms are well-documented in specialist marketing books. Virtually any campaign can be defined, structured, and have its cost-effectiveness assessed by using that template of headings. Avoid using agencies, contractors, or marketing communications executives who are prepared to provide services and material *without* demanding such a framework.

Many campaigns work better when mechanisms are combined. A good example is the use of telemarketing to follow up direct mail shots. Another example is the use of a special offer at a trade show. At the appropriate point, the lead is handed over to the sales team.

But in the case of the WWW, the situation may be significantly different. Since intermarketing is interactive, the customer will respond to being given a route map of her own to take her to the point of placing an order. Take the following example:

Motorcycling enthusiasts may be attracted to a Web site that contains information on particular makes or models—and even a sponsored bulletin board. Having browsed around the site, the prospect looks at a special offer of software that will provide a maintenance planning schedule. It's free for a month and he/she downloads it and brings it into use. Three weeks on he/she is mailed: "Do you want to go on using the program? If so, then please register on our database and we'll send you a five year license, still free."

The opportunity is now there to offer paid for upgrades and extensions to the software, and also to e-mail-shot details of bikes, parts, and holiday sales.

The Web site also contains an online order form.

The service provider is setting things up once; it is the customer is doing most of the work thereafter.

Attracting the customer to the Web site in the first place is the hard part. Imagine the benefit and the payback of communicating with existing customers online.

7.6 MEASURING PERCEPTIONS—MANAGING STAKEHOLDER DIALOGUE

7.6.1 Measurement

An essential part of campaign management is measurement. This takes many forms: direct mail items sent, inquiries received, Web site *hits,* take up of special offers, leads generated, and cost of campaign. But there is a saying in competitive business: "perception is reality," so the marketing function must not only measure marketing activity, but must also identify and measure perceptions and how they are shifting.

If customers are to be gained and then kept, then their opinions must be sought and acted upon. It is exciting to develop and execute marketing campaigns that spread the name and gain new customers, but for a business that depends for profit upon maintaining long-term customer relationships, it may be even more important to look after the customers that already exist.

Chapter 11 covers this area in detail—including the measurement of implementation and operational performance. The role of marketing is to measure the softer issues. In Section 11.7, it is recommended that simpler is better. Just four questions will elicit 80% of all that is there to be found out and by virtue of focus and open question format, that 80% will provide 125% worth of actionable value:

1. What do we do well?
2. What would you like to see us do better?
3. What has changed over the last six months?
4. Will you or anyone else you know want to buy more of our product in the near future?

This simple formula is adaptable. An employee attitude question set would be the following.

1. What do we do well?
2. What would you like to see us do better?
3. What has changed over the last six months?
4. Can you think of anyone who should be moved within or brought into our team?

Questions can be adapted for all four groups of stakeholder: customer, employee, stockholder, supplier. The task of the CEO is to give each group what it wants. In practice, this implies managing the expectations of each in line with reality. If there is dialogue between groups of stakeholders, each with its own knowledge resources and paradigm, then priorities will be easier to assess and the "reality" may move for each party. Effective management of this dialogue will make the difference between a perception of success and a perception of failure.

7.6.2 Stakeholder Dialogue

The marketing function can help in a number of ways as well as measuring perception. There is also the framing of messages, management of events, publication of newsletters, and so forth. The marketing/inter-stakeholder dialogue needs to consolidate many viewpoints, as shown in Table 7.2.

Table 7.2
Perceptions and Sources

Constituency	*Sources of Perception*
What customers perceive about service	May be incomplete and 6 months out of date
What front-line employees perceive about providing a good service	Is probably based upon today
What senior managers perceive about what they are doing about today's issues	May be based on plans already formed
What stockholders perceive about the safety of their investment and the prospects for steady growth	Comes from all kinds of sources and through filters....including the fluctuations of the share price
What the public perceives	Is what is read in the papers or seen on TV
What customers' friends perceive	May define the difference between success or failure in the longer term.
What dialogue ensures is...	A sharing of reality, priorities, and a common base of time

No telecommunications-based service provider will succeed if it is not adaptive.

Dialogue between customer and firm is not just a question of *aligning expectations*. Satisfying or even exceeding expectations is not sufficient. A second, powerful, outcome is that dialogue will increase everyone's knowledge, or store of useful information so that they do something about it—enhancing a product, buying a new feature, giving a favorable press comment, and so forth. A crucial point is that customers often leave because they didn't think anyone cared about them once they'd placed an order and paid their first fee. Conducting an explicit program of dialogue will mitigate this problem.

Wherever possible, employees and contractors should have the same access to information and they should have the same care taken in strategic dialogue as anyone else. It is said that firms with good public attitude ratings are one and the same with those that have high employee attitude ratings.

All stakeholders should benefit as the enterprise as a whole learns, adapts, and keeps moving ahead of those enterprises that do not.

The following activities are examples of mechanisms for strategic dialogue:

- Brand advertising;
- Public relations: press releases, journalist briefings;
- User group—both meeting physically and on the Internet;
- Internet bulletin board and answers to frequently asked questions;
- Stockholder and investment community briefings;
- Quarterly reports and mandatory information filings;
- Help line and help screen;
- Seminars, including opportunities to meet senior managers and professional specialists;
- Visits to facilities;
- "Customer focus groups;"
- Preferential business terms;
- Sponsorship;
- Newsletter;
- Employee briefings—team briefings;
- Joint initiatives in total quality management.

7.6.3 When Everything Goes Wrong...

There should be a crisis plan prepared in advance for every aspect of the organization's processes. The status of previous and current stockholder dialogue will generally have a key part to play. Success in containing the presentational risks will, to a large degree, depend upon how well the corporation has built up its networks, its stock of goodwill, and of credibility.

It is an unfortunate truism that disaster fascinates people. It can also bring out the best in the people who have to fix things. But it is the way that problems, events, and recovery plans are communicated that may make the difference between overall success and failure. The time before a crisis arises should be used to build up a stock of goodwill (and trust) and to develop contingency plans. Thinking through how to handle what may go wrong will, in any case, contribute to thinking on how to prevent problems in the first place.

Disaster can strike in any of a multitude of ways:

- Someone *hacks* into the system and makes unauthorized transactions on a customer account, ordering services from an information provider;
- The call center burns down and service falters as the backup plan comes into action.
- The stock price plummets 30% on rumors that a major customer contract is going to be lost;
- There is a threat of suit for product liability;
- A key employee leaves and sues the firm.

The examples show a spread of stakeholders—customers, suppliers, investors, employees. All will need to be informed and reassured that their interests are being protected. If a crisis hits, then the following checklist will help:

- Set up PR function within crisis management office and have all media calls directed there.
- Appoint a spokesperson and assemble facts of the case.
- Brief senior management and agree on a strategy to clear up and to improve for the future.
- Ensure that all stakeholders who need to be briefed are briefed (NB stock exchanges' rules on disclosure of information).
- Brief trusted media contacts and assess likely interest and points of sensitivity.
- Take part in crisis management meetings, advising on PR aspects and remaining well briefed.
- Continue briefings and handling calls until the crisis ends.

7.7 INFORMATION MANAGEMENT

7.7.1 The Process and its Relevance to Marketing

Soon, information management will be defined regularly as a process area in its own right. At present, it might have been defined as a process within practically any of the eight areas defined for a telecommunications-based service provider. Each process area has a responsibility for managing its own knowledge.

There is a particular emphasis in marketing—it is there that market research is collected, and it is within this function that strategic dialogue of stakeholders is managed. Marketing/intermarketing is a gatekeeper of information.

7.7.2 Internet Librarian and Webmaster

There always has been a need for people who could identify and navigate around information resources and bring back the information required for business decisions. Librarians and market researchers have existed for generations. But there were barriers to their use: Librarians sat in libraries that were somewhere else, and market researchers cost money that had to be found from somebody's budget or signing authority. Large firms had their own information sections—normally run as an overhead.

The advent of the Internet as a front-line business tool has changed the paradigm. Librarians and specialist researchers have been using the Internet for years, but now everyone wants to be able to do so. Not everyone can.

If possible, an Internet librarian should be identified who will assist in the following ways:

- Help employees connect and navigate for simple inquiries;
- Show how more complex searches can be carried out;
- Carry out inquiries/research talks;
- Identify the key information resources needed by the firm;
- Maintain, in several media forms, all necessary information on competitors, key customers, and suppliers;
- Maintain a media watch for important and relevant information.

The Internet librarian may also be the *Webmaster* who will manage the creation and maintenance of the firm's Web site to the required standards and formats.

These new roles are a combination of the technical, marketing specialist, coach, and much else besides.

7.7.3 Decision Support Systems—The Marketing Data Warehouse

The analysis of customer usage of products and individual features, functionality, or arrangements in particular was discussed in the context of developing a new product concept (see Section 8.11.2). The marketing data warehouse is no more than a store for data that enables the use of detailed and long-term client histories and of trend information.

Airlines, already identified as telecommunications-based service providers rather than simply as operators of flying machinery, are near the forefront of work in this area. Coming from an environment of nonintegrated, stand-alone systems and databases, airlines are now clear in their intent to focus on the activities and behavior of their customers in order to manage the relationship with them for increased revenue, customer satisfaction, and profit.

With the information available in its systems, an airline is able to do micromarketing to support transition from order taker to active seller. Building out of basic data such as seat and meal preferences, airlines are able to treat each customer as a valued individual if the relevant information is presented to the screen as soon as he calls the service center for information or help.

Product planning is improved as a result of identifying functionality that is used and that which is not. Cost-benefit analysis can be carried out. The product developer can concentrate on improving commonly used features rather than rarely used ones.

Essentially, all the information imaginable is stored on relational databases. Data on customers, products, pricing, channels, costs, and other factors are all recovered from the *data warehouse* and analyzed in combination using a massively parallel mainframe computer. The results of such analysis provide information to business management on the best combinations of variables to maximize profit for the factors of production deployed and under the constraints that exist. Online operational systems are partitioned from data warehouses, so they are not slowed.

Figure 7.3 illustrates how this all works for a bank.

7.8 SUMMARY

In the telecommunications-based service provision environment, the rate of change and complexity is high; the scope for misunderstanding is high; the importance of keeping people aligned is high. There is a continuous need for communication, for dialogue, and for sharing of information. However, it is not sensible to simply share every piece of information with every person. Just as telecommunications-based service provider customers look for simplification and relevance in the product and service offering, so do the stakeholders look for a common direction and a shared perception of events and performance.

One of the outputs of the process is information that will lead to the creation and improvement of products—the subject of Chapter 8.

It is reasonable to assume a key task of marketing/intermarketing is to achieve recall of one's own corporation within, say, the top three

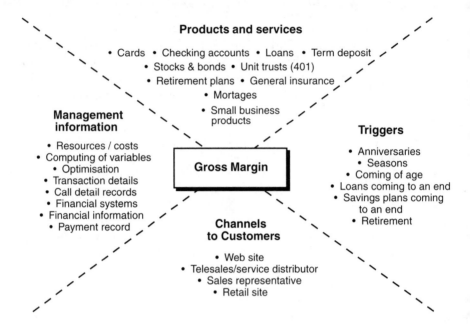

Figure 7.3 The "market of one" from using customer information.

names associated with the particular niche and a successful retention of customers thereby obtained. Success in this regard is likely to come from four things:

- Ruthless focus on market, however defined, and *precise* value proposition;
- Use of mechanisms in combination;
- Absolute commitment to consistency of message;
- Backing of a well-designed and delivered product, experienced positively by existing customers.

Products

8

8.1 OVERVIEW

Chambers Dictionary defines a product as "a thing produced—brought into being...producing value...a result...quantity got by multiplying." The term *a product* is frequently synonymous with *a service* in the telecommunications arena and the terms are frequently interchanged in this book.

The traditional methods of creating and managing products are insufficient, without enhancement, for the service provider's task of bringing complex, short-lived information and technological products to a tightly defined customer base in the new political, economic, social, and technological environment. This chapter will do the following:

- Identify these additional considerations;
- Provide an enhanced architecture and tools for service provider product creation;
- Describe the underlying process, tools, and systems for development, launch, and subsequent operation.

The creation and maintenance of products implies the management of each product through its life cycle of conception, trial, development, launch, in-service support, postlaunch enhancements, special promotions, and product withdrawal. Product management includes all of the volume, pricing, and costing aspects. Why is it that in the new environment of telecommunications-based service provision, there are new factors to be considered and wider architectural approaches to be taken?

It is difficult to overemphasize the significance of billing and customer support systems in the overall scheme of a particular product for a telecommunications-based service provider. For convenience, design issues are discussed together with operational issues in Section 12.9 and in Chapter 14.

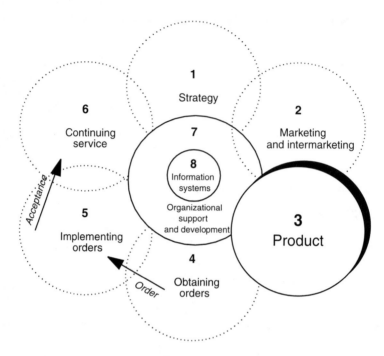

Figure 8.1　Process area 3—creation and management of products.

8.2　ADDITIONAL CONSIDERATIONS

Firstly, the breakdown of national and industry boundaries discussed earlier and the continuous interaction and interdependence of business practice with breakthrough technology requires a broader understanding of the *politico-economic* environment than has formerly been the case. For example, the satellite delivery of feature films to cable TV head-ends (the receiving point for signals distributed around a cable network) was inhibited in Europe by country-by-country content-related performance licensing arrangements conceived for the reel-to-reel technology. Governments around the world have varied in their enthusiasm for an open competitive telecommunications environment, so that what is allowed in one place may still be difficult to achieve in another. Those who would put together products in the new environment need the knowledge and sensitivity to identify and understand the traditional barriers and cultural perspectives and the creativity, determination, and leadership to remove the barriers and exploit the greatly increased opportunities thereby exposed.

Second, there are now more complex *socio-technical* factors—the demands of customers and the technology that can be brought to meet them. Traditional product management approaches were quite sufficient to meet the demands of product development under conditions of broadly homogenous markets, relatively long product life cycles, and single technology. But those methods will probably not be agile enough or be wide enough to cope with the increased speed of development and the significantly increased demands of customers, and the methods need careful enhancement if they are to embrace the complexities of mixed/multiple technologies.

8.3 PROCESS REQUIREMENTS

Therefore, the telecommunications-based service provision approach has to be one that takes a determined direction towards the (moving) ideal of each customer, but a more flexible approach to the technical methods used at each stage. To achieve this three components are needed:

1. Clear direction and linkage with the other processes and actors in the telecommunications-based service provider (the purpose and leitmotiv of this book);
2. Constant architectural principles;
3. Tools and disciplined processes to build and manage product element within the architecture.

From the *Chambers Dictionary* definitions and from the discussion in Chapter 4, telecommunications-based service provision products and services in general would seem to be "links from customers to elements and combined elements of information and technology," so that they can obtain business benefits and improve quality of life (or perhaps more accurately "get more out of their lifestyle"). But, in the fast-moving competitive arena of telecommunications-based service provision, the established methods have been accelerated, adapted, and linked more strongly to the increased demands of customers for low prices, individual features, and immediate and easy use. There are increasing expectations that products and services will be universally and globally available, without break, 24 hours a day.

8.4 PROCESS DESCRIPTION

What emerges is a composite process of the established product development and management methodology, informed and amended by special telecommunications-based service provision considerations and possibilities. Architecture is likely to be based around the principles of the call center, computer telephony integration, and the Internet. The complex technical content of telecommunications-based products demands a sound *information systems design* approach. The high rates of change inmarket and technology indicate a greater need for *maneuverability* and *iterative prototyping* rather than for extended product development cycles.

It may be appropriate where possible to anticipate developments that are not strictly feasible today and essentially provide for what will be available in two years time by functional bridging in the meantime. This is described in Section 8.11.

Table 8.1
Process Outline—Creation and Management of Products

Field	Attribute
Process area description	Create and manage product
Process owner	Product marketing
Processes and activities	Concept (prototype may occur here) Business case and technical feasibility Design Prototype, trials, and development Launch Sale, delivery, and operation In-service development Withdrawal
Time to complete	From 1 to 12 months to launch Increasing possibilities in "online configuration" move this period downwards Variable product service time
Inputs (and sources)	Business plan objectives ("strategy") Ideas to *create new* products (various) Ideas to *improve prices and functionality of existing products* (various, but particularly via stakeholder dialogue—see Section 7.6.2)
Outputs (and destinations)	New and existing products to sell ("marketing/intermarketing," "obtain orders") Changes to products already in service with specific customers ("provide continuing service")

Field	Attribute
Lead user for these information systems	Product: All internal and external Information and physical elements bought in and/or supplied to customers Product Information, generic terms, conditions and pricing, configuration, etc. Product support systems: order input, billing, etc.
Support documents (Includes interactive documents, forms, etc. on home page)	Product creation, development, design, and specification documents Product brochures and other publicity material Order forms, etc.
Effectiveness measurement (measured by those receiving the outputs)	Market sizing Benefits/value to customer, comparison with competitive offerings Orders and market share achieved Time to market and delivery time Ease of sale, installation, and operation Customer satisfaction Margin *Acid test:* Present value of gross margin achieved (N. B. same as "obtain orders")
Risk and cost management/ disaster recovery	Risk: Getting it wrong for demand, salability, competitiveness, timing The product fails to gain a "position" in the customers mind (see Section 7.4) The product does not work properly or is too expensive in resources Product liability Costs: Components, processing, sale, distribution

8.5 ARCHITECTURE—AS IT RELATES TO PRODUCT CREATION AND MANAGEMENT

This chapter is devoted to the specialist considerations of product design and operation. The information systems architecture that underpins a telecommunications-based service provider as a whole is given in Chapter 14.

Architectures provide a taxonomy, unchanging framework, and vocabulary in which individual elements can be developed separately but deployed together.

There are at least five elements that provide further frameworks for thought and action to enhance the use of the established subprocesses described in Sections 8.6 to 8.10. These architectural considerations are as follows:

- OSI stack;
- Call-center-based entity model;
- Systems development life cycle;
- Maneuvers;
- Similarity to the major bids process;
- Interim bridging.

8.6 ARCHITECTURE—THE OPEN SYSTEMS INTEGRATION STACK (SEVEN-LAYER MODEL)

It is generally held that information systems should operate to common and *open standards,* whereby the components from one manufacturer will interwork with the components of another. The benefit for the users of products are that they can choose from the widest possible range of products to meet their overall needs. Initially, when you bought a computer, you were also confining yourself to the narrow range of software that worked upon it.

Consider the difficulties customers meet when the plugs on the power leads of the items they buy do not fit the power supply outputs on the wall and when the voltages are different. It may well be too late to standardize the electrical power characteristics of the nations of the world but much of telephony and computing has already been effectively aligned. Bodies such as CCITT, ANSI, and ETSI have already worked for years to this end. In the narrower world of computing, standards have also derived from market dominance—consider Microsoft, Apple, Intel, Oracle, or from a mixture of the two. Notwithstanding, there will always be a very large number of apparatuses, protocols, and signaling standards that are incompatible. Similarly, the use of language between humans in different countries and the use of language between humans and machines remain areas where there is still more difference than standardization. The resolution of these differences, in specific cases, is one of the tasks of the service provider.

The use of layer models, or stacks, is helpful in illustrating the hierarchy and relationship of different levels of the technologies that are brought together to make products. All items within a stack must have the same characteristics (e.g., the position of the pins in an RS232 interface plug). The boundary between layers describes the way that, for example, an RS232 plug would be used by the layer above.

The CCITT seven-layer model, or open systems stack, was designed to enable the development of advanced telecommunications and is described in Table 8.2.

Table 8.2
Open Systems Integration—Seven-Layer Telecommunications Model

Layer	Content
7. Application	User applications and services that support them
6. Presentation	Syntax to enable applications to cooperate with each other
5. Session	Parameters and methods of exchange between presentation facilities, from connection to disconnection
4. Transport	Use of communication resources to transport data efficiently between cooperating session facilities
3. Network	Handles routing of data within or between transport facilities
2. Link	Controls transmission of data across a link, including error checking and recovery
1. Physical	Mechanical and electrical interfaces (e.g., pin positions, plug sizes, etc.)

8.7 ARCHITECTURE—CALL-CENTER-BASED ENTITY MODEL

Products cannot be considered in technical terms alone. To ensure their alignment with strategy, with other products in the portfolio, and with existing/common processes, it is important to consider products within the architecture of the firm as a whole. The very design of this book places "creation and management of products" within an organizational architecture designed to show the relationship of *processes* around a *timeline* or as an overall cycle. What is needed, also, is a description of *product* within the other *entities* of the firm.

But, if "a company is its products," then how, semantically, can this be achieved?

Telecommunications-based service provision is likely to be based around a call center. It can be helpful to consider the call center as the central entity around which the telecommunications-based service provider is built, as diagrammed in Table 8.3. As in any *architecture stack,* the purpose of the diagram is to show the hierarchy and interrelationship of the individual elements. Many of these are directly recognizable, too, from the value chain in Fig 4.3. The overall information systems concept is developed further in Section 13.5. The details of call centers are covered in Section 9.6.

Table 8.3
Service Provision—Based Around a Call Center

Element or Entity	*Description*
Company direction and administration	Creation and monitoring of strategy Administration and finance, decision support
Customer Relations	Marketing, sales and relationship management, project management, especially through...
Call center or similar	...(Outbound)proactive marketing, sales, or customer satisfaction activity (Inbound) Customer call reception, customer record call-up, taking action or transferring caller to others, monitoring to completion or resolution, clearing call from customer for...
Customer care	...Processes of installation, billing, fault rectification, upgrades, user help relating to...
Products and services	Customer facing information applications...
Integrated platforms	...With common look and feel to the customer and with effective transaction processing through client-server/hub & spoke/online transaction processing, switches, store-and-forward devices, etc. with...
Technology elements	...Network, computing, terminal equipment, software elements

8.8 ARCHITECTURE—SYSTEMS DEVELOPMENT LIFE CYCLE

The development of products for telecommunications-based service provision is, almost invariably, a complex technical task. As discussed in Section 10.10, telecommunications-based service providers are very similar to systems integrators. The processes of product development and management should therefore take account of the systems development methodologies that they use. Whatever product development methodology is used, it should be cross-checked against these stages/work breakdown structure components.

The following components of a typical method are largely self-explanatory:

- Establishing user needs;
- System design;
- Implementation design;
- System component integration;
- Installation;
- Operation and training;
- Program management—commercial;
- Performance monitoring.

8.9 ARCHITECTURE—MANEUVERS

The concept and practice of *maneuvers* is discussed fully in Section 7.9. They provide a framework for the rapid deployment of products in response to opportunities. In summary a maneuver has four components:

- Spot the opportunity;
- Implement it quickly;
- Take insurance about it going wrong;
- Improve the solution.

Good maneuvers have the following characteristics:

- They are cash-positive.
- They are simple and probably quick.
- They move you in an appropriate direction to a better market position.
- They are within your area of knowledge and focus.

- They are not confusing to anyone other than your competitors.
- They don't prevent you from executing a better maneuver.
- They can be executed better and faster by you than by your competitors.
- They have containable risk—a fallback solution.
- They don't compromise your core business.

As well as providing a method for rapid moves, the maneuver checklist works well as a checklist for any telecommunications-based service provider product; it is particularly appropriate for a board/senior management review. Prior to using the maneuver approach, there must be a clearly articulated vision of the world ahead and the mission of the firm within that vision.

8.10 SIMILARITY TO MAJOR BIDS PROCESS

It is sometimes said that a major bid is "a new product for a single customer." A major bid is usually carried out quickly; it equates to the maneuver in that many of the considerations above apply. The major bid process comprises the following, described more fully in Chapter 10:

- Finding the opportunity;
- Being invited to bid;
- Assessing the opportunity...go/no go decision;
- Creating the solution;
- Preparing and delivering the bid document;
- Negotiating and closing the sale.

As the product environment moves more and more to the satisfaction of individual needs (the "market segment of one"), so do the product management methods look more like the major bid process. Common sense dictates that all products must make money—the go/no go decision for changes and enhancements to the underlying product is crucial.

8.11 ARCHITECTURE—INTERIM BRIDGING

For most practical purposes, anything is technologically possible. If the need is expressed properly, then a technical solution can be developed. It is frequently said that we are limited more by our imaginations than by the technology.

Technology evolves when the need is strong. People will use difficult technologies if the payoff is high. Who complained more about the sound quality and call drop-out of mobile phones, business users or private users? For the former, it was a valuable business tool with a high utility and often paid for by someone else. For the latter, it was an expensive social device, inferior in quality to the fixed phone. The latter complained more—unless the mobile phone was being used to summon the fire brigade to their burning house or an ambulance and paramedic to a dying child at a road accident.

To start with, technology may be clumsy and expensive but, as the three following examples illustrate, it can get better integrated, easier to use, and cheaper as it goes along.

8.11.1 Example 1—Clumsy

There is a wonderful scene from an old Bob Hope spy movie that illustrates this point. Bob Hope is setting out on his mission, taking with him a pen-sized global communications device that has just been given to him with eulogies about its performance. Unfortunately, he also has to take with him the two heavy suitcases containing the necessary batteries. But, 30 years later aren't we are on the verge of having the pen-sized device on its own? For years, too, sci-fi pieces have taken the voice-computer interface for granted. We can do that now. In a few more years, we'll do it all of the time.

8.11.2 Example 2—Difficult To Use

Electronic mail started as a combination of analog fixed phone line, external modem, and early PC with a rudimentary, nonintuitive human/computer interface. Connectivity between networks and systems was poor. It was a good system for techies; other people learned to love the fax machine instead.

However, the systems are getting better and more integrated. You can, for example, now use a Windows-based mail system on a laptop connected to a LAN, the Internet or straight through to the GSM networks. Fax is now integrating with the PC. Graphics can be incorporated in or attached to electronic mail documents. The number of e-mail users is now increasing fast, and it is really becoming worthwhile to get connected. Finally, electronic commerce is a reality. But, if the introduction of e-mail had been delayed until everything about it was considered easy to use and the market had been fully defined, then it would never have been introduced at all.

8.11.3 Example 3—Expensive

One of the world's leading telephone-based banks has steadily moved its customer transaction processes from the retail branch to the telephone. The bank is an innovative user of online transaction processing, interworking client servers linked to legacy systems, card systems, and interactive voice response. A key objective is to maintain its progress towards phone/computer and away from paper, but the technology for one aspect, new password registration, was not yet available in a sufficiently secure form. Rather than retreat from the overall strategy, the bank opted to create a subprocess whereby new customers carried out the entire registration with one agent and for the final subprocess were transferred to one of several agents in an electronically shielded, locked room who asked one single question of the new customer whose identity they did not know: "What password do you wish to have?" "CATKIN" "Thank you, goodbye." Clumsy? Yes. Expensive? Yes. But...in line with the key product marketing objective of providing a one-stop/single transaction paperless process? Indeed, yes. The technology will soon be available to do this automatically and less expensively.

In the fast-moving arena of telecommunications-based service provision, it is better to introduce a product with interim/bridging aspects than to miss the window of strategic opportunity and lose the customer.

8.12 SUBPROCESS 1—CONCEPT

8.12.1 Introduction

There are plenty of books dedicated to product management, and most of the subprocesses will be described in bullet form only. The conception of products in telecommunications-based service provision, however, warrants a fuller treatment.

Telecommunications-based service providers come in various forms (as discussed in Section 4.12). Whatever their form, however, service providers' products have the following characteristics:

- Information and technological elements bought from suppliers (or supplied from within the same enterprise in the case of the third form of telecommunications-based service provision) and resold to end customers; adding value, such as advice, tariffing, billing or equipment supply (e.g., cellular service provider, video hire shop, credit card company cross-selling elements from a third party).

- Information or technological elements that are *switched* or *reprocessed* before reselling them (e.g., Internet service provider, international/ long distance telephony resellers).

Service providers operate under a range of strategic regimes (see Chapter 5). Some T-BSPs will be niche suppliers, others may be low-cost suppliers. But all must identify and introduce new products or product enhancements, so that they are not trapped with low margins as the existing product offering becomes commoditized or is made obsolete by the introduction of a better substitute.

Taking the example of the mobile service provider, the range of opportunities addressed rarely moved outside the single dimension of mobile telephony. That is where the money was being made originally. Fixed-mobile convergence created combinations of mobile phone call and private circuit, thus giving substantial price savings to those with corporate private networks. The convergence of mobile telephony and datacoms created mobile data—now given a real boost by the spread of GSM. Call forwarding functionality on PABX equipment and on digital public exchanges created a form of universal personal telephony. But the business and technological opportunities created by convergence are not defined within just mobile and fixed telephony. Practically any need, once identified and described, can now be met with a combination of technologies that can be refined and simplified over time.

Breakthroughs can be made when you begin to use technologies in combination.

Consider for example the following combinations: *global positioning system* (GPS) linked to a microcomputer; mobile telephony with electronic mail; data broadcasting with CD-ROM; intelligent networking with computer integrated telephony with relational databases of customers; and in-flight entertainment systems.

8.12.2 Product Invention by Structured Brainstorming

As an exercise, imagine creating technological capability to support some service requirement that draws upon a combination of items on the sample list of capability found in Table 8.4 from mobile and fixed telephony, added to computing technology and information management. You can work with items from just one column, but work especially with ideas from the first linked with ideas from the second and/or third column.

Examples of such combinations include the following:

Table 8.4
Combining Technologies To Create New Products

Computing/ Information Management	Fixed Telephony	Mobile Telephony
Call handling for third parties	Messaging systems	Mobile data
Call center	Leased lines	Mobile telephone handset
Internet access service provision	PABX—fixed/cordless	Satellite-based systems
Internet Web page authoring	Voice processing systems (e.g.	interactive voice response)
Computer telephony integration (CTI)	Call logging equipment	Laptop PC with PCMIA card
Multimedia work station	Fax machine	Paging
Outsourcing	Long distance resale	Call records
		Road traffic reports
Hub and spoke architectures		Road toll booth equipment
Database management systems		Traffic radar systems
JAVA script to create applications on Internet	Free phone numbers (800)	Mobile fax
Billing system	Directory services	Car radio
Web browser		
Video	CD-ROM	

- *Credit card authorization plus mobile phone:* This combination is already in place and enables us to book theater tickets or to teleshop while on the move.
- *Hub and spoke architectures plus interactive voice response plus CTI plus road toll booth equipment:* This might enable you to tell a road toll booth to charge your bank account from your bank account. It would probably be easier to use a debit card, of course...
- The reuse on a computer terminal of interactive voice response applications/dialogue structures originally developed for voice applications might enable the rapid and easy provision to the public of information held on a WWW database.

However, the business and technological opportunities created by the convergence of these technologies are still only part of the picture. Add convergence with "entertainment," "information services," and "education and culture." Imagine adding a further column into the bank of capabilities in Table 8.5:

- Feature film;
- Arcade game;
- Chess;
- Star Trek;
- Museum opening times;
- Train fares and times;
- Holiday reservation service;
- Text book;
- Pictures of Roman remains at Ephesus;
- Live pop concert;
- Plumbing services;
- Paying for a one-year parking permit;
- Consulting a doctor.

It is relatively straightforward to capitalize on the inherent capability of intelligent networking and tailor marketing and business strategies accordingly. This can be done on a small scale or a large scale. The convergence of "mobile," "fixed," "computing and information management," entertainment, information services," "education and culture," "shopping," "banking," and whatever else one chooses, provides a multidimensional opportunity—with complexity to match.

The objective is to arrive at a new combination of *desires and possibilities, applications and technologies* to open a new market or to improve what is there already.

There are plenty of ways to develop these ideas in practice. Some depend mainly upon the intuitive "right brain" and some upon the more deterministic processes of the left.

8.12.3 Concept Development

Here is a set of ideas and methods...

- Have technologists brainstorm combinations of technologies, as illustrated above, to create new technical possibilities that can be applied to existing problems: "Let's use IVR to interrogate the Internet...what happens when you put together the concepts of text-to-voice, online language translation, and the CD-ROM on a really powerful UNIX box?"
- Have customers muse about their dearest wishes, not only for the products that are both familiar to them and exist, but also for prod-

uct concepts that do not: "If only the mobile phone could...if only my bank would..."

- Ask suppliers and potential suppliers what they would like to see as new markets or uses for their products—but bear in mind that questions they ask may contain presumed solutions, so the questions should be probed for the real need: "Could we sell these theater tickets over the Internet?" perhaps should be rephrased more directly as "Can you think of additional channels to market or different packaging possibilities?"
- Brainstorm *dissimilar* concepts to arrive at totally new combinations: *vegetarian sausage.*

Some of the resulting concepts will push the bounds of credibility, will be before their time, or will be considered absurd in scale. How must people have scoffed at the concept of *science fiction.* The expression is an oxymoron; the words conflict with each other.

A recently seen poster in Frankfurt airport recalled an earlier prediction "There will never be a market for more than 1,000 cars in Europe." Remember the famous saying about the telephone: "I can see no great use for this device because the supply will never dry up of small boys to run messages." In the face of another similar perception: "...one day every town will have a telephone," how hard it must have been to conceive universal personal telephony. It was probably Star Trek that showed us the possibility in its first convincing form.

But it is dangerous to introduce ideas before their time, for the marketplace will not embrace them or the technology, despite all efforts, which may be too rudimentary or unreliable. Similarly, the conditions in one market place may not be replicated elsewhere. What is taken as a normal process in one place may be totally unknown in another. What will work in a country with a well-developed banking law may be totally impractical in another that is not so blessed.

Harking back to Chapter 7, it pays to remember that slogans and humor rarely travel from one culture to another.

8.12.4 Formal Market Research

This could include deterministic approaches, from desk research on new markets for existing products, perhaps from direct feedback on product performance, or customer complaints, or from clearly emerging demands. Direct comparison of competitive offerings will yield important information on required functionality, price, performance, and customer service. Market research is a key purpose of stakeholder dialogue, described in Section 7.6.2.

The following ideas list may help:

- Desk research;
- News clippings;
- Becoming a customer;
- Becoming a shareholder;
- Trying out demonstrations;
- Visiting competitors' WWW sites;
- Visiting competitors at trade shows;
- Asking your own customers about the competition—but it is normally better to avoid reminding them that they are there at all;
- Appointing one person in the company to "be" one of your competitors—to track all that they do, to build dummy proposals for bids that you are doing, to identify strong and weak points, and to identify business that they *must* win to forward their strategic intent.

8.12.5 Product Concepts From Decision Support Systems

A further breakthrough in product management derives from what can be analyzed, inferred, and simulated from previous transactions held in the firm's *data warehouse.* Here can be recorded everything a customer does, every inquiry made, every preference expressed, every transaction carried out (even down to each telephone call).

The recent improvements in database management systems (e.g., Informix and Oracle parallel architectures for relational databases) and massive capability in parallel processing make it possible to analyze all of this data with each variable considered against another. The resulting absolute and trend information can give rise to precisely targeted products. Consider the following examples.

- Example of SAGA—a holiday company for the over 50s now evolving into a financial services company...using its highly valuable customer database.
- Telcos can tell you which calls you made most frequently and invite you to group these numbers in a discounted frequent calling plan. In practice, privacy and citizen's rights issues mean the telco is unlikely to emphasize this capability or to share it with other people.
- Banks can call you on the birthday of a child to offer particular financial services, or just before your 25th wedding anniversary to ask if you had ever thought of borrowing money for "that trip of a lifetime."

8.12.6 Environmental Changes

New laws or deregulation, the opening of a new rail link, changes in attitudes to debt and borrowing are all examples of changes that could affect your business, both for the good and for the bad.

8.12.7 Documenting the Concept

From one or more of the above approaches will come a product concept that will find its way to the next stage. The new concept should be documented to describe its purpose, market, underlying technology, and so on. It may be that the concept can be demonstrated as a prototype; perhaps a simulation can be constructed. Sometimes a new concept can be explained as a scenario product—as a story, a multimedia presentation, perhaps.

8.13 SUBPROCESS 2—BUSINESS CASE AND TECHNICAL FEASIBILITY

In constructing the business case, there will first be an analysis of the opportunity, a measurement of market size, and a costing of the development and of the product in service. Pricing models will be prepared and assessed. The technical approaches will be discussed and subjected to close scrutiny.

Those who assess new products know that the inventors will enthuse, the relevant business unit or account manager will declare the product to be one of strategic necessity, and the sponsors will underestimate the costs and timescales. The technical inventor may have overestimated the capability of the technology proposed; it may be untested; the concept probably takes several existing components and assembles them; the new combination may have hidden incompatibilities.

Conversely, the sponsors may have underconsidered the breadth and magnitude of an opportunity; technicians may have failed to spot much more obvious ways of doing things or may have been trapped inside an existing paradigm.

It is likely that decisions to take product concepts to the next stage of development will be made by people from several parts of the organization—it is certainly better so to do, for it ensures balance and overall knowledge. The team assessing the new idea will be provided with a checklist of considerations. Many of these have been identified in the sections above. It is not worth going further to define them here.

Has the concept been considered by people from just *one* organization? Given the increasing convergence of technologies and businesses, it is imperative that a broad range of knowledge, skills, and perspectives is brought to play.

The bottom-line considerations are as follows:

- Does this new (or improved) product take us towards our strategic goals?
- Is the market planning realistic? Have the sums been done properly?
- Can we afford to do it? What will happen if we don't (...and our competitors do...)?
- Are we genuinely adding value to the elements we are buying in?

With information services products, there is a further consideration. The real gains come when money services and financial transactions are included. The explosive growth of globally accepted credit cards has been a major enabler of teleshopping—with more transaction methods like electronic purses coming onstream fast; the cross-border phone bank without an expensive branch network is a forthcoming development. An information service without a booking capability is an increasingly unlikely concept.

If a concept is accepted, albeit after amendment, then the next task is to create the design of the product.

8.14 SUBPROCESS 3—DESIGN

The design of a product proceeds through the stages below. In each case some examples are given of the activities or questions concerned.

- *Statement of the requirement (Why?):* What the customer wants to achieve and at what cost. Service products are used by business customers to "improve their competitiveness" and by private customers to "get more out of life." What price can we charge and how many will we sell at each price?
- *Logical design of the product (What?):* What technical components will it comprise? How will they work together? From where will we source these components—should we make them for ourselves? How much will the components cost? What are the technical design risks—obsolescence, dead-ends, nonacceptance of technology (e.g., VCRs ended up being developed to the VCS standard even though the Betacom standard was widely preferred.) How will we bill this product? What will be the user interface?
- *Implementation design (How and When?):* How long will it take to produce the product? How much will it cost to bring it to the point

of launch? Do we need to do alpha and beta trials with selected customers or our own employees? How will we integrate the product components? How will we deliver/install the product? How will the sales force sell it (see Table 9.2)? How will we train customers in its use? How will we document the product? Which billing system are we actually going to use....will it be ready in time?

- *Performance monitoring:* How long do we expect this product to be in service? What shall we do to phase its introduction? How shall we measure what it does—for the customer and for us?
- *Pricing:* There was an approach to this in Chapter 5. Pricing could be the subject of a book in its own right—and will be found in any good work on market planning. There are no immediate issues in telecommunications-based service provision that are not inherent elsewhere. Pricing is either derived from the costs of the elements or from what the market will bear in competitive terms, from what the owners of the firm wish to see as margin, or as a direct function of the value of the service in the value chain of the buyer. Pricing is the art of arriving at answers that make these congruent.

For a telecommunications-based service provider, the bulk of the issues will be around the presentation and deployment of the product to the customers and around the selection of suppliers rather than relating to the construction of the product elements themselves. Many of the product performance issues will be contained within the product elements that are bought in complete from strategic partners. In Section 6.7, the issues of strategic partnership are considered in more detail.

The consideration of how to sell the product is a particularly large component of the product design subprocess of the telecommunications-based service provider. It will be appreciated that the telecommunications-based service provider is itself a channel to market for the suppliers of the individual technical and information elements. Table 8.5 serves as a further aide memoir on many of the aspects of product development. A full discussion of sales force design and of third-party channels to market is given in Chapter 9.

8.15 SUBPROCESS 4—PROTOTYPE, TRIALS, AND DEVELOPMENT

The content of this subprocess is described within the title. The possibility of doing trials was considered in subprocess 4 (Section 8.12).

Prototypes and dummy products should be made available as early as possible. So many products have silly faults that could have been

Table 8.5
Aide Memoir for an Internet Service Provider Product

	Volume	*Niche(s)*	*Large*
Market Size, structure, and importance	10m home and small business PC users Some very high users	Academic community Dealer networks of PCs and similar items	Any medium-sized or large corporation without its own "co" connection to the Internet
Product	Commoditized: access software and popular services	As for volume customers plus: Access to thesis database and student helpline and discount services Bulletin board	product helpline, technical information, stock/spares listing, and ordering mechanism Brochure distribution
Spatial location and density of customers	100,000 customers, evenly spread throughout	10,000 students and faculty at university campuses and institutes. Especially strong at University X 500 Local dealers of XYZ personal computers Mostly urban (more in north than in south)	500 large customers with 50 to 200 desktops 50 customers with 200 to 1,000 desktops 10 customers with 1,000 plus desktops including one with 12,000 desktops
Time of day	Mainly day and evening—some all night/24 hour users	As for volume	As for volume—peak at 0900 to 1030 when people first log on
Product knowledge requirements	Need to coach new and infrequent users in basic functions and provide access to more complex features and services	As for volume— plus special services for each niche Requires access to detailed information on XYZ products and company procedures	As for volume— with some skilled library/market research capability In "desktop management mode" need capability to receive and prediagnose problems of network, hardware, software, and user capability.

Table 8.5 (Continued)

	Volume	Niche(s)	Large
Customer knowledge requirements	None special	Nothing special Need to understand needs, environment, and pressures	Need to know individual company procedures and critical commercial issues
Price	Revenue per customer	$500 per annum Commoditized and Highly competitive	$300—high usage at discounted price $1,500—normal services at normal rates plus $1,000 at high margin for special services
Cost of sale/service level requirement	Minimum over high volumes consistent with achievement of basic service levels	Minimum cost and service level acceptable. Academic users will prefer to affiliate with like-minded agents "Instant response" required during dealer opening hours	Need to set up deals properly in the first place and to run normal services at lowest cost relative to contracted level

eliminated at an early stage if the users could have had a chance to use them. Also, there may well be accessories to be produced.

In one instance, a dummy handheld terminal was found to be just too big for a standard case on the open market; a request by the telecommunications-based service provider to the manufacturer of the terminal for a minor change to its dimensions eliminated the extra cost of having a carrying case made specially.

During the development stage, there will be a particular need for careful strategic dialogue—listening, sharing, and telling the market and other stakeholders what is coming. At the right time, the media must be briefed. Dialogue will reveal which aspects of the new product are going to create the greatest level of interest.

8.16 SUBPROCESS 5—LAUNCH

No aspect of launching a new product that is specific to a telecommunications-based service provider product comes to mind, though there are many matters in general to be considered. These matters come under the two general headings of impact and efficiency.

On *impact*, it is said that one of the most important aspects of achieving impact is timing. There will always be a degree of luck associated with the media coverage that will be given on that day. Consider the following two examples.

- Pity the product manager whose new project was brought to the attention of the public the day the Gulf War began.
- On the other hand, the U.K. launch of the Mercury One-2-One PCS network in 1993 was on a quiet day for news and made the first item on the national news; or, rather, it was the promise of free calls that caught the interest.

Leaving aside the luck factor, the important thing is to have booked space for advertising, to have primed the media in advance, and to have identified the points in which there will be interest. To catch interest otherwise and to obtain sales, there is a whole armory of devices from balloons to free airplane trips, from "money off special launch offer" to free demonstrations/upgrades for existing customers. It may be that the telecommunications-based service provider wants to give a party to mark the day and to thank those concerned, but the general concept of a launch party is subject to challenge. The announcement of a large immediate order, complete with a generous endorsement can do little other than good.

Probably the best approach is to retain a good marketing agency to do the whole job for you.

Efficiency is a matter of having everything about the product ready, employees trained, brochures and order forms available, customers lined up, stocks lined up, and so on.

8.17 SUBPROCESS 6—PRODUCT SALE, DELIVERY, AND OPERATION

See Chapters 9, 10, and 11.

8.18 SUBPROCESS 7—IN-SERVICE DEVELOPMENT

The in-service development of products is often considered less interesting and important than the development of new ones. In terms of immediate margin, this is unlikely to be the case. A customer already won is an excellent prospect for the sale of improvements to that which is already working reasonably well.

Not only is it normally more profitable to sell to an existing customer, but the added opportunity for dialogue will keep the telecommunications-based service provider in the mind of the customer. Finally, as outlined in Chapter 12, customers who have a complaint or problem that is handled well can become even more loyal than before.

In-service developments may be iterative improvements, new functionality, a higher grade of service, spares, accessories, and add-ons. Another high margin, easily sold enhancement is training.

Given these factors, it is evidently worth putting a specific focus on sales of items to in-service customers. If sales and marketing are more inclined to focus on new orders, then appoint a business manager with bottom-line responsibility for the former. Margin and customer satisfaction will grow.

In managing the improvements, patches, and rectifications required from subcontractors—the suppliers of individual technical and information elements—it is essential to work through a single interface and to manage required progress against an *open items* list.

If a part of the telecommunications-based service provision product is critically dependent upon supplied software, then be sure to have the supplier agree to put source code *and* all subsequent patches, new releases, and associated documentation into escrow (that is, held by a specialist third party, against the risk of the supplier going into liquidation). Be sure to actually try out the current escrow-held version from time to time.

8.19 SUBPROCESS 8—WITHDRAWAL

For most products, there comes a time when the telecommunications-based service provider should stop to sell them and then to remove them from service. It is a matter of increasing costs or of reducing margins vis a vis new products now available. Furthermore, a customer with an old product is at risk from the blandishments of competitors comparing their new with your old.

A special offer to customers with obsolete devices or outmoded information services will frequently be the most effective way of taking old products out of service and, as in Section 8.15 above, getting a new product off to a good start. Similarly, a special offer of a middle-aged product might solve the old product problem and relieve an overblown stock situation. For a telecommunications-based service provider in a fast-moving and complex market, it is vital to retain focus. It may be helpful to sell off the spares and source code (but not the customer database) and outsource service and liabilities to another organization or, say, to a group of former employees. Whatever is done, the long-standing customer should be approached as valued, loyal, and deserving of considerate attention and preferential treatment.

8.20 RISK, COST, AND DISASTER RECOVERY

This has already been covered in the process outline shown in Table 8.1, but...

Check, check, and check again that the billing system meets the requirement and operates well. Telecommunications-based service provider products may well be centered upon the call center, but their effectiveness and prosperity is most vitally affected by the effectiveness of the billing system.

Be absolutely certain that the customer database is backed up, protected from intruders, and that it can be deployed instantly in the event of a move to a secondary site.

Keep asking the question: Why should somebody be buying this product and service rather than buying one from the competition or using a substitute?

8.21 CONCLUSIONS

Telecommunications-based service providers exist in a fast-moving, competitive environment where industries and technologies converge and overlap and where the combined effects of globalization and deregulation also act. Given the "instant" nature of telecommunications-based communications, there is a clear need to enhance the marketing models that have served well in the 1970s and 1980s but were, perhaps, designed around cars, furniture, and soft drinks. But the rules and processes developed for those slightly earlier times and less frenetic environments do give a good basis.

The creation and management of products in a telecommunications-based service provider is a mixture of conceptual, contractual, technical, and practical aspects to bring to market the technical and information elements of others to an ever more critical set of customers. There has always been a need to consider products within the political/economic/social/technical (PEST) environment analysis and to position and enhance them according to the strengths/weaknesses/opportunities/threats (SWOT) assessment. This remains. It has just all become faster and faster.

This chapter, together with those on strategy, market communication, and obtaining sales, has examined modes of thought that can be added to the traditional. As in every aspect of the telecommunications-based service provider corporate entity, there is a need for products to be managed by a multidisciplinary team centered upon a technical marketing core. By the combined use of technical architectures as shown and traditional methods, the innovative and thorough service provider should be able to deploy products that will match any need and create powerful market positions.

Obtaining Orders

9.1 OVERVIEW

Only a minority of enterprises are effectively exploiting the opportunities to sell and manage external relationships with telephone-based, information system supported methods. Those that do are benefiting from more and better sales at lower costs because they have reduced the time for the process, greatly improved precision and accuracy, and lowered personnel and other costs.

The processes of marketing, including advertising, direct mail, telemarketing, and, increasingly, the Web site, are designed to bring a flow of leads from targeted prospects to be converted into profitable orders by the sales force. (See Figure 9.1.)

The successful conduct of the sales process in a telecommunications-based service provider requires a mixture of flair, determination, knowledge, and discipline. Sales management requires a high level of attention to detail and a keen use of teamwork.

9.2 PROCESS DESCRIPTION

The purpose of this chapter is to describe the activities that have to be carried out in obtaining orders for the products the service provider has created so that they may be installed and invoiced with excellence. In every case, there is an underlying sequence of events: preparation, inquiry, presentation, close, and administration. The chapter is arranged to consider this underlying process in respect of volume, specialist/niche, and large customers ("major accounts"), which are all handled in different ways in terms of both scale and of character.

As before, the starting point is a formal description of the process, as shown in Table 9.1.

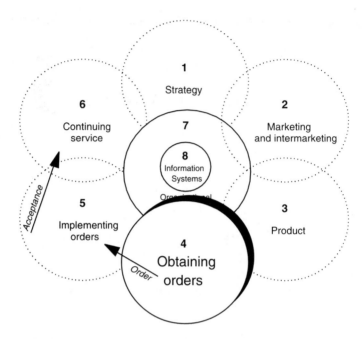

Figure 9.1 Process area 4—obtaining orders.

Table 9.1
Process Outline—Obtaining Orders

Field	Attribute
Process area description	Obtain orders
Process area owner	Sales
Processes and activities	Classic sales process comprises: Preparation Inquiry Presentation Close Administration Processes and channels vary by customer type: large
Time to complete	1 to 12 months
Inputs (and sources)	Business plan objectives ("strategy") Products
Outputs (and destinations)	Agreed and documented sale to initiate the "implement orders" process

Field	Attribute
Lead user for these information systems	Outbound telephony: telemarketing—scripts, predictive dialers, etc. Selling systems: "prospect to order" tracking, bid preparation, sales management
Support documents	Account development plan Customer contact records Prospect lists Customer order documents including: Customer requirement specification System design Implementation specification
Effectiveness measurement (measured by those receiving the outputs)	1. Number and quality of opportunities identified 2. Times "short-listed" 3. Orders achieved 4. Ratios of 1
Risk and cost management/disaster recovery	Risk: Monitor sales "pipeline" to ensure opportunities are exploited and obstacles to success are removed Insist on "team approach" to each opportunity and documentation of activity Ensure you sell what you can deliver within planned timescales and costs. Insist on early input from implementation team Manage pricing and contract terms Monitor sales team focus and effect of incentives (strictly congruent with strategic objectives) Recognize sales effort and success right across organization *Costs:* Use appropriate channels and sales resources to provide resources/tools in line with opportunity Ensure correct level of sales and product training Review sales targets and incentives frequently *Disaster recovery:* Emphasis on prevention, not cure—especially on pricing, deliverables, terms and conditions

9.3 THE CLASSIC SALES PROCESS

9.3.1 Relevance of the Process

The classic sales process phases—preparation, inquiry, presentation, close, and administration—are as relevant to telecommunications-based service provision as to any other enterprise. The phases occur in a major sale of a complex computer system; they occur in the sale of a television set in a shop; they occur for all of the sales carried out by a telecommunications-based service provider, whether for volume, specialist, or large customers, and there are particular issues of communication to be considered when customer and salesperson are not communicating face-to-face.

Nearly every interaction with a customer or potential customer can be treated like a *sale*. All interactions provide an opportunity to provide information and service, to negotiate in some way, or simply make the customer happy to be there.

9.3.2 Preparation

Following the marketing processes that bring the product to the point where it is offered for sale, a suitably sized sales team *prepares* by being in the right place at the right time with the right training, information, tools, and support to receive inquiries from potential customers and then conduct the sale through to its ultimate conclusion. A prudent salesperson precompletes sales order forms to save time later.

9.3.3 Inquiry

Although customers will often know precisely what they want and will ask for it straight away, they will generally start by making a general *inquiry* about the product or service on offer. It is important at this stage to structure the sales conversation so as to learn the precise customer need rather than launch into a description of what you have to sell. You don't want the customer either to end up ordering something that is unsuitable or to terminate the sales conversation because you have confused or bored him or her or because you have allowed misconceptions to persist. A properly trained salesperson will concentrate all early efforts in eliciting information rather than giving it out. The salesperson is not passive in the process but should be encouraging the prospective customer to articulate his or her need accurately and thoroughly, to consider new possibilities and, above all, state the value of satisfying each aspect of the need. Some needs are psychologically based rather than tangible. The activity is complete when both parties are agreed on what the customer needs and on the benefits that are required.

For the telecommunications based sales process, there are opportunities to prompt both the salesperson and the customer. Screen prompts can help. Case-based reasoning software will help an agent to step through the customer inquiry process in a logical manner. Interactive voice response (and Internet equivalents) allow the same.

9.3.4 Presentation

Salespeople then *present* their solution to the needs. If they have been alert and methodical in the preceding stages and can recall and relay the relevant aspects of the products they have to sell, then they can present the solution in the terms of what their customers have said they want to buy. It is perfectly legitimate to place more emphasize on those aspects that match the needs well. It is also legitimate to minimize the aspects that do not match so well and suggest ways that shortcomings might be overcome. Where a product has features or shortcomings that are not relevant to customers' needs, then they should not even be mentioned. It is not ethical, sensible, or (in many countries) legal to pretend that a product will meet a specified need when you know full well that it will not. At each stage, the salesperson checks that the customer understands what has been said and that each aspect of the solution will meet her need. Where the customer raises an "objection" (i.e., says that the solution will not provide the required functionality), then the salesperson must find ways of redefining the solution or reducing the scope or importance of the requirement until the customer is satisfied. Where the customer raises a price objection, then the salesperson must define the price or, for example, premium over another product from another supplier in terms that relate directly back to the value to be obtained by the customer. *Value* is given if a customer is able to do more than before or is now able to pay less for the same as before. In each case, the process of overcoming objections should be continued until the customer expresses herself satisfied with the information given.

The presentation of a product or solution entirely by means of a real-time telephone conversation can be extraordinarily difficult. Customers may want to take time to consider or they may want to feel or see what they are going to get. Objections and buying signals will be more difficult to detect without access to the nuances of direct, nonverbal signals such as body language. The customer and salesperson may be in different countries or time zones; they may be using an unfamiliar language; understanding may be impaired by cultural differences.

Similarly, a sales presentation conducted over the Internet—through a Web page—or by a visit to a shopping mall is interactive but only insofar as the structure and programming have predicted and been enabled *in advance.*

It will often be helpful to combine methods of communication, one with another.

One cable TV company teleshopping company displays its goods as a television program and customers connect to a call center to make their purchases. The retailer presents its IVR method as a positive method of jumping to the front of the line by using self-service

9.3.5 Close

The *close* comes when the salesperson believes that the customer is ready to agree. There are many other books that address the subject of closing a deal; this is a brief summary and note of the factors relevant to service provision. Until the presentation has finished the salesperson is encouraged to avoid conflict or tension. Many salespeople are bad at asking for the order and closing the deal. Some, of course, are predatory by nature and revel in the hunt. Many use their high energy levels to create large numbers of sales encounters and thus more opportunities to close deals. But many people fail entirely to ask the customer right out for what they want. The salesperson has had a good measure of control, but now control shifts. It is the customer who decides whether or not to buy. Many salespeople are frightened of giving over this control and, if they are affiliative by nature, they may actually fear rejection. In many cultures we are schooled to avoid asking for things outright.

Examples of closing techniques include the following:

- *Direct:* "I seem to have answered all of your queries, Mrs. Smith. Can I go ahead and complete the contract?"
- *Alternatives:* "Have you now decided whether it will be the red or the green one that you will have?"
- *Indirect:* "What day would you want delivery?" "How will you want to pay?"
- *Multistage/intermediate:* "It seems we should arrange a free demonstration?" "OK...I'll send you the brochure, Dr. Mitchell."
- *Time-bound:* "Joan, if you order before the 30th, I can offer you a 10% discount. If you don't order now, we'll probably run out of stock."
- *Puppy dog close:* "Please, will you hold this adorable puppy dog for a few minutes?"

An effective telephone salesperson will probably deploy all of these at some time or another, depending upon circumstance. Clearly, given the difficulty and limitations of communicating across a limited bandwidth (i.e., just voice or just using e-mail) should lead you to consider scaling down what you are asking from your customer. You cannot sell an electricity generating station over the telephone, but you can *sell* an appointment with one of your specialists. You cannot sell a car over the telephone, but you can *close* someone by asking them to take a free test drive.

The Internet provides an excellent system for "trying before buying" when a prospective customer is invited to download software to use on a license for either a limited time or for limited functionality. The follow-up for a close is made much easier when the customer has committed that far in advance (similar to the *puppy dog* close).

9.3.6 Administration

Having gone to all of the trouble and expense of closing the sale and getting the order, it is essential to thoroughly complete the *administration* tasks that follow. Order documentation must be completed (thus translating the once *fuzzy,* but now well understood, requirement of the customer into the precise *rectangular* terms of the product configuration), billing database records created or amended, delivery schedules checked, and stores drawn and dispatched.

The documentation of these aspects will normally be defined by the sales order and include a clear description of the customer requirement, a system design, an implementation plan, and the information required to bill the customer.

In practice, the first two may be defined by the packaged product being sold. The supplier has a duty to ensure that the customer's requirement, as stated, will be met by the product being offered. To what extent this is achievable depends on both parties. Where the requirement is vague (either in fact or in the way that it is articulated) or where it can be seen that the product will fall short of meeting what it is clearly required, then this should be made clear. The precise legal position of the supplier in this regard will vary from country to country.

The telecommunications-based service provider should arrange its systems and processes so as to take down all of the required information at the time of the call or Web site visit.

A telecommunications-base event booking service will normally combine its close with its administration: The details of event and credit card are recorded and then repeated back. Then the agent says: "You can collect the tickets from box office window 11 on the night at any time after 7 p.m. Once you have put the phone down, you will not be able to cancel these tickets."

There is always one other thing to be done in telecommunications-based service provision or any other form of business transaction. It is to thank the customer for the order and to reassure them that they have chosen supplier and product well.

9.4 CHANNELS TO MARKET

The underlying questions of channel management are well-understood and covered in sales and marketing texts. However, the fundamental changes brought about by the telecommunications age have only recently begun to be documented. Many organizations are missing the opportunities and advantages that are now available.

Channels to market have proliferated as market segmentation has become more important and other factors such as cost of sale have come under increased scrutiny. New services such as the Internet have also made sure that channels to market have, in some circumstances, moved from professional delivery to consumer/home delivery, and in the case of the Internet and mobile communications, specifically to a lifestyle type of channelization. Consider these examples: 24-hour banking for international travelers; lifestyle brands reaching out to their database of customers to provide a new service; low-use consumer packages/tariffs for mobile telecommunications; Internet seen as a hip/cool lifestyle choice rather than boring old e-mail, and audiotex chat lines.

Part of the very concept of service provision arises from the requirement to sell and service products and services through effective channels to market. Telecommunications-based service providers *are* a channel for the suppliers of technological and information elements. Telecommunications-based service providers are in three fundamental forms as covered in Section 4.12.

- In the *first form*, the SP buys information and technological elements from suppliers and resells them to end customers, adding value such as advice, tariffing, billing, or equipment supply (e.g., cellular service provider, video hire shop, credit card company cross-selling elements from a third party).
- In the *second form*, the SP also *switches* or otherwise processes the information or technological elements before reselling them (e.g., Internet service provider, international/long distance telephony resellers).
- In the first two forms, the assumption is that the SP and the element providers are independent of/external to each other and that their relationship is defined by that factor. A *third form* of service

provider is an *internal service provider* (e.g., customer sales and service helplines and airline reservations lines).

All forms of telecommunications-based service providers face the issues of channel strategy. In all cases they must select, create, and operate the appropriate channels through which to market the products thus available to them.

It would be unrealistic and wasteful for every customer to have at her side 24 hours a day a personal salesperson with total knowledge of both her product and of her customer's business or lifestyle. The market channel decision is, therefore, the resolution of the following factors:

- Market size, structure, and importance;
- Spatial location and distribution of customers and suppliers;
- Time of day;
- Level of knowledge about the product;
- Level of knowledge about the customer;
- Price and revenue per customer;
- Cost of sale.

Analysis of those issues will enable decisions on when and how, for example, to deploy

- Salespeople who physically call on customers;
- Retail outlets;
- Telecommunications-based sales people;
- Customer self-service, including automated voice response;
- Mail-based sales techniques;
- Electronic mail and Internet-based techniques;
- Relationship specialists, technical specialists, market specialists;
- Other intermediaries—distributors/dealers, systems integrators.

The following table gives an example of the analysis that could be carried out.

Table 9.2 illustrates a first cut channel analysis and plan for an Internet service provider supplying access software and value added information services into a single national market. The scope of the analysis overflows the "obtaining orders" process into implementation and service issues.

The example was chosen to demonstrate the range of customers who might be addressed with economy by judicious reuse of the same elements, albeit with varying emphasis, configuration, and added functions for some.

Table 9.2
Channel Analysis and Plan: Internet Service Provider

	Volume	*Niche(s)*	*Large*
Market Size, structure and importance	10m home and small business PC users Some very high users	Academic community Dealer networks of PCs and similar items	Any medium sized or large corporation without its own ".co" connection to the Internet
Product	Commoditized: access software and popular services	As for volume customers plus: Access to thesis database and student helpline and discount services Bulletin board, product helpline, technical information, stock/spares listing and ordering mechanism. Brochure distribution	E-mail and bulletin board services. Access to business and specialist databases—including a market research service and "specialist newswatch" service. Option to manage desktop services
Spatial location and density of customers	100,000 customers, evenly spread throughout	10,000 students and faculty at university campuses and institutes. Especially strong at University X. 500 Local dealers of XYZ personal computers Mostly urban. More in north than in south	500 large customers with 50 to 200 desktops 50 customers with 200 to 1,000 desktops 10 customers with 1,000 plus desktops including one with 12,000 desktops
Time of day	Mainly day and evening—some all night/24 hour users	As for volume	As for volume—peak at 0900 to 1030 when people first log on

	Volume	Niche(s)	Large
Product knowledge requirements	Need to coach new and infrequent users in basic functions and provide access to more complex features and services	As for volume—plus special services for each niche Requires access to detailed information on XYZ products and company procedures	As for volume—with some skilled library/market research capability In "desktop management mode," need capability to receive and pre-diagnose problems of network, hardware, software, and user capability
Customer knowledge requirements	None special	1. Nothing special 2. Need to understand needs, environment and pressures	Need to know individual company procedures—and critical commercial issues.
Price, revenue per customer	$500 per annum. Commoditized and highly competitive	$300—high usage at discounted price $1,500 Normal services at normal rates plus $1,000 at high margin for special services	Competitive discounting at time of winning business. Low rates for usage but opportunity to charge realistically for help-desk services or for consultancy/integration
Cost of Sale/service level requirement	Minimum over high volumes consistent with achievement of basic service levels	Minimum cost and service level acceptable. Academic users will prefer to affiliate with like-minded agents "Instant response" required during dealer opening hours	Need to set up deals properly in the first place and to run normal services at lowest cost relative to contracted level
Primary channel to market	Outbound sales: brochure and sign-up forms with outbound telesales follow-up	As for volume market but... Use students themselves as sales people and as call center agents—	Initial sales made by specialist sales force calling on customer. Continuing service managed by nominated individuals and

Table 9.2 (Continued)

	Volume	Niche(s)	Large
Primary channel to market (continued)	Specialist help desk to activate, connect, and assist. Call center to manage billing and service N. B. special "Gold Service" for high users	partly paid by highly preferential usage vouchers Short response times and quick transfer with "screen popping"* to back office help on XYZ products	teams—using same information tools as for other markets but tailoring these to each major customer's requirements Regular progress/monitoring meetings of project teams and senior managers from ISP and customer
Secondary channels and support aspects	Interactive response and access to ISP databases (e.g. for account information, customer detail update, etc., both by voice and PC to a home page	As for volume market. Student user group, self-help groups and "discount club" organized on campus basis.	As for volume market Service management, technical and contractual specialists
	Regular newsheet with hints and tips new product information etc.	EDI links to XYZ factory and to brochure distribution point. Person to visit customers on regular basis.	Indirect sales may take place through Systems Integrators
	Indirect sales may take place via dealers, clubs, and societies (possibly using an "own brand") etc. Indirect sales may also occur by having the ISP service bundled in with the sale of a PC—banking service, computer equipment, etc.		

* Screen popping: computer telephony integration whereby the contents of an agent's screen are transferred to the screen of a second agent simultaneously with the transfer of the caller to whom the screen data refers.

It needs little development of the model above to see how the channels for a global information service for, say, SCUBA divers or veterinarians might be provided. The channel strategy would reflect the need for the service provider to keep its products at the leading edge of thought in those selected sectors. Reflecting on questions addressed in Chapter 8, one would expect to see representatives of the service provider at industry conventions and to see them contributing to the market sector written media. Furthermore, the agents handling customers in such specialist markets would need to be conversant in issues, priorities, and technical terms and probably able to provide service in several languages. The combination of low international call charges, the functionality of the Internet, and the happy fact that a 24-hour service becomes ever more viable as you add more time zones into the market's geography mean that the utility to be gained by divers and veterinarians will easily exceed the costs of providing them with the service. There would be advantages in appointing geographically local distributors that could supplement, face-to-face, the transaction-intensive activities of the call-center-based agents.

But the real gains come when money services and financial transactions are included. The explosive growth of globally accepted credit cards has been a major enabler. The multinational bank is a logical development. Cross-border *direct banking* in the single European market is now regarded as an inevitable development. Such an idea again looks so good because of the low costs.

Roughly 6% of European consumers did their banking by telephone during 1994, and the market is expected to grow to $1,000 billion by 2000. (*European Banker,* April 1996). But the issue is not just about domestic banking. The fact is that all banking markets will use telecommunications-based services for the transactions that it serves best. One of Europe's leading bankers predicts that virtually all balance inquiries and simple account transactions will eventually be made via interactive voice response or networked PC inquiry mechanisms. This enables the banks to both become markedly more cost-efficient for all the transaction-intensive part of their work and newly enabled to concentrate on customer relationship management and value added product sales. There will still need to be bank staff meeting larger customers in person and carrying out particularly complex or sensitive business face-to-face.

The lesson is that there are no longer simple relationships between single channels and single-market segments. It is the creative deployment of more than one mechanism that will define the future.

There is an implication for assigning the management responsibilities within a firm. In the Internet service provider example shown in Table 9.3 there is no *sales director/VP* with overall responsibility for all

Table 9.3
Assignment of Management Responsibilities in an Internet SP

Title	Customers	Obtain Order Processes	Other Processes
VP	Customer operations	Volume, student niche	Outbound and inbound call center sales, help desk operations, interactive voice response, and home page access to databases
VP, Major Customers	Major customers and market sectors	Relationship management and direct sales force management	Relationship management and quality of service throughout the telecommunications-based service provider value chain. Providing specific customer-dedicated support teams within the call center environment
VP, Indirect Channels	Dealers, "own brand" resellers, and those bundling in ISP, systems integrators, XYZ chain, and similar	Distributor development and revenue generation. Relationships with desktop management suppliers and consultants. Marketing materials dispatch	Distributor support and problem resolution
VP, Sales Support	None, but provides bid managers as required	Sales engineering, special configurations, demonstration facilities, training, bid management	Skills/resources/ training for help desk

sales or even all sales processes. There are four such people who all re-port directly to the chief executive officer.

As can be inferred from the table, the four VPs handle the channels that primarily utilize the processes for which they are also responsible or call upon specialized knowledge that resides with them and they call upon each other to provide the resources and process execution for which each has responsibility.

For example: the VP, major customers, manages the relationships with the large universities, but the volume product directed at students is managed and deployed by the VP, customer operations. The VP, sales support, provides support, knowledge, and resources to the VP, major customers, as required. The VP, customer operations, is not precluded from having some direct sales or support people to visit customers; however, it is more likely that such services would be available more cheaply from one of her colleagues.

In each case, it is the interplay of the process and customer responsibility that defines the best place to put accountability of one or both.

From the analysis and examples in the sections above, it should be possible to design and operate a simple or combination channel strategy to address any situation. The next three sections check off some of the issues.

9.5 VOLUME SALES PROCESS—CALL CENTERS

9.5.1 Positioning of the Call Center

The concept and role of the call center was considered in the Introduction and then touched upon in several preceding sections and many of the examples. In particular, in Section 8.7 it was portrayed as the pivot of the processes and resources deployed to handle virtually all of the volume customers and, as shown above, many of the simple/repeated transactions for other sections of the market, such as major customers. Refer again to Table 8.3.

9.5.2 Components of the Call Center

To deploy such a call center philosophy requires most or all of the following components, shown first together in Figure 9.2 and then described individually.

- *Automatic call distribution (ACD):* Provides automatic extension of calls to the next available agent according to predetermined routing rules. The ACD also provides performance measurement and statistics to help determine manning levels and individual agent

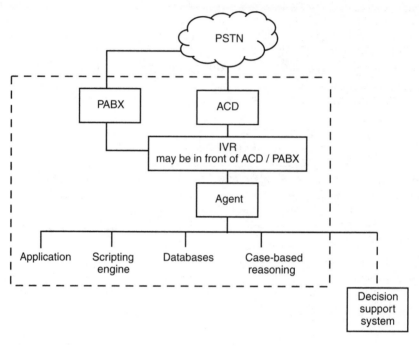

Figure 9.2 Call center architecture and components.

effectiveness. In most call centers, the number of calls waiting, mean time to answer, number of calls answered in the last hour, and so on are displayed for all to see.

- *Interactive voice response (IVR)* (alternatively AVR, automatic voice response or VPS, voice processing system): Provides prerouting of calls to the appropriate functional area and at a more sophisticated level enables the caller to use the telephone as a terminal device attached to a host database. Possible steps involved in such an interaction are listed in the following.

  1. System answers incoming call: "Good morning. This is Midtown Bank. For Sales, dial 1. For Billing and Customer Service, dial 2. For other inquiries, dial 3."
  2. Caller dials "2."
  3. System answers: "Please enter your account number, then * your PIN and * again. If you do not know your account number, dial 3."
  4. Caller dials "989898*2211*."

5. System answers: "You have entered nine eight nine eight nine eight. For balance, dial 1. For recent transactions, dial 2. If you have another inquiry, dial 3."
6. Caller dials "1."
7. System instructs host: Read balance field for account 989898 and output result as speech.
8. System speaks: "Your balance is four thousand five hundred and twenty-two dollars and fifteen cents. To end, dial 1, for more options, dial 2. To speak to an operator, dial 3."
9. Caller dials "1."
10. System answers: Thank you, good-bye."

- *Call center applications:* Written on a specialist server or on the IVR server to handle the regular or specific campaign functions designated for the call center.
- *Scripting engine:* Call centers will generally provide scripts to their agents as part of inbound and outbound applications.
- *Switch:* The switch receives the call from the public switched telephone network or an incoming private circuit and, in conjunction with the ACD, extends the call as required.
- *Databases:* These hold the data that populates the applications.
- *Computer telephony integration (CTI):* Links together the voice and data components of a call. Enables the *screen popping* function described in the footnote to Table 9.2.
- *Data warehouse/decision support system:* Holds historical or very rarely used data away from the online databases, thus enhancing the response of operational *real-time* transactions. As already mentioned in Chapter 9, there are huge benefits to be obtained from detailed analysis of customer transaction data leading to improved decisions on product features and market segmentation (see Section 7.7.2 and Figure 13.2.

9.5.3 Web Site as Call Center

In addition, it should be remembered that *callers* may be interacting with a database by use of their own PC. They may be networked via the Internet or other data network. They will probably interact via a Web site.

In the mid 1990s, few mass market service providers were considering the Internet as a (physical) channel to market upon which they would focus their limited management attention and information systems resources.

However, several factors have acted to change that perspective:

- Niche market considerations—students and "Wall Street types" alike use the Internet every day;
- Natural development of the home pages that many firms now consider to be an essential part of their marketing communications mix;
- Continued pressures to reduce call center labor costs by automating functions and tasks;
- Increased appreciation of the benefits of combined media strategies (i.e., the use of data rather than voice, of Web page interaction rather than agent interaction enables a more leisurely and detailed browse for the caller);
- The spread of *intranet* technology—becoming the information systems architecture of choice for many firms;
- Realization that IVR applications can be ported simply and directly to become Web site interactive devices (see Figure 2.3).

There is a saying that "call centers are the sweatshops of the 90s," though hardly a call center like the modern and attractive one shown in Figure 9.3. Call centers can give an impression of lots of people handling lots of uncooperative customers and boring transactions in an unthinking and identical fashion. There has certainly been a *danger* that call centers have not been satisfying places to work, but the tendency is passing. Monotony is also receding in the measure that "mass marketing is dead" and we are moving to the more interesting demands of "markets of one" and to "external management."

Here are some of the reasons why and how call centers are getting better:

- Applications are becoming smarter.
- The capacity to cover basic transactions by automated methods means that the agent is given more interesting inquiries to handle.
- IVR, with or without voice recognition, is becoming rapidly more prevalent and more customer-friendly.
- Reducing costs of manufacture are enabling greater focus on niche markets and more resources in an organization are directed at the customer interface.
- Database management systems and decision support techniques are providing additional information that enhances campaigns and improves customer perceptions of the offerings or responses they are getting. ("Ah, Mr. Jones...I see that you normally have a Grand Circle seat. Would you like two in row M at $45?").
- It is possible to identify the most profitable customers and select certain agents to provide them with a "gold star" for more personal service.

Figure 9.3 Photograph of a call center.

- Experience with call centers has created a body of knowledge about the environmental and motivational factors that make for success. The topic of staff motivation is on the agenda of many call center conferences.

There have been many decades of attention to the processes and automation of manufacturing. It has only been in the last few years that "white collar" work has been given the same attention. There is still a long way to go in most places and industries.

There are wide national differences in the attractiveness and acceptability of call center work—in Singapore, for example, where educational standards and expectations are high, it is said that call center work is regarded as demeaning. Conversely, there are economically depressed (and therefore cheaper) areas in most countries where the work is gratefully welcomed. The combination of cheap telephony and regional variations in labor costs and other variables give rise to a study in their own right.

Call centers appear in the remodeled factories and warehouses of earlier industrial eras. Awards are earned by and given to the architects of these attractive and efficient new places of work. Groups of call centers spring up in the same towns or industrial sites. Staff turnover can be high.

9.6   HIGH-VALUE CUSTOMERS AND NICHE MARKETS

Higher value customers may be so defined for a number of reasons:

- High-volume users of basic services;
- Users of high-price options and services (e.g., specific databases);
- Customers in specialist knowledge niches;
- Major customers—telephone contact is efficient for less complex matters;
- Individuals who are high-value by virtue of their position or influence (e.g., senior executives of major customer accounts, albeit acting in their personal capacity; their relations; politicians; journalists; celebrities).

These customers need to be handled in accordance with the precepts above, but the service provider needs to go a step further in making them feel valued and well cared for. They may well be routed to the most experienced agents. In the case of *specialist knowledge niches* (the examples of veterinarians and SCUBA divers were given earlier), they should be handled by people who themselves also have that specialist knowledge and understanding. These people may be handled in priority call queues. We may take time to call them proactively to a regular *calling plan*. The high-value customers will probably be handled by named people or by a small number of people in a dedicated team. A generic term for these people who will be able to develop longer term profitable relationships with their customers will be *telephone account managers*.

Telephone account managers and highly skilled sales people are now able to make quite complex sales over the telephone. They are highly trained and they are supported by agile and well thought out *contact management* screen-based applications. Because they are so dependent upon these applications, they tend to become highly developed and often may provide a higher level of quality for the customer than a salesperson using standard "pen and pad" ordering. Furthermore, there is an enormously increased capability to retrieve detailed computer-held product information.

Once again, telephone account managers and sales people will often be encouraged to go out and meet their customers face-to-face—or will be supported by people who can do that for them.

Before leaving this subject, it should be mentioned that the assistance of a call center agent may be essential for people who have health or ability impairments and are not able to use the normal applications being offered to them. Students in the United States are required to register for their courses at the start of each semester. Many academic institutions have automated this process using IVR and associated techniques. The institutions are required to provide access to all; it pays them to send an intermediary to their homes or to have specially trained agents to assist them.

9.7 MANAGING MAJOR ACCOUNTS

9.7.1 Relevance of Telecommunications-Based Methods to Large Customer Sales

A service provider, like any other business, would certainly devote resources to managing large business customers with account managers and specialists who take time to learn all about their customers' businesses and to develop close professional relationships with their executives and technical staffs. Nonetheless, as already stated, large customers may have routine processes that can be handled more efficiently by telecommunications-based methods, including electronic data interchange (EDI) and call centers/telecommunication account management than by people calling in person.

Account managers of large customers should be based upon a call center. That is where the customer files should reside.

One of the key roles of the major account manager is to create and maintain an *account development plan*. Who of the traveling account manager and the fixed and computer-supported agent is better placed to gather and maintain information such as customer annual reports, press cuttings, sales brochures, and telephone lists?

The increased tendency for major account people to be location-independent and equipped with mobile telephony and voice messaging creates, perversely, an even greater need for support by well-trained and well-equipped base staff. On the whole, this is not done. Major account teams are typically starved of support resources. Such resources are labeled as *admin* and *overhead*. They are spread across several sales people and often have little knowledge of the company's customers. In offices with standard telephony configurations, they will not be able to perform their vital functions as effectively as agents equipped with the appropriate support tools and information, as already described.

Furthermore, if account managers bear continuing responsibility for quality of service (as surely they should), then the service functions of billing and maintenance will need to be handled to a high standard by the individuals customers reach when they phone in their problems.

9.7.2 What Characterizes Major Customers

Major customers may be the top 10, 20, or 50. Whatever the case, a major customer will be one who provides a significant amount of revenue and/or is highly influential in providing a significant amount of revenue from elsewhere.

In each case, a major customer will be one for whom there is a named account team and for whom there is an account development plan/customer service plan. A major customer will also be one for whom there will be major bids from time to time.

9.7.3 Account Team

The account team is responsible for the overall business relationship with a major customer. In particular, the account team *hunts* for new opportunities and *farms* the customer service relationship that already exists. Depending upon size and importance, the account team will comprise the account manager (otherwise called account director or account executive) managing the overall process systems engineer who advises on account technical strategy and generates system solutions, project managers working on specific implementations (see Chapter 10), and operational/customer service personnel (see Chapter 11). Sales to the customer will be carried out by the account manager with or without product specialist sales people. Specialists from contracts, accounting, training, and so forth may be formally assigned to specific accounts. There should be a senior manager who *champions* the account and maintains senior level relationships.

9.7.4 Account Development Plan/Customer Service Plan

The documentation of the account relationship is holistic and continuous, including, as it should, specific details of services being supplied.

The account plan is designed to support and document the processes and plans for increasing revenues from new orders and building upon existing business. Within the documentary architecture of the account plan will be all of the information about customers: individuals, teams, and plans. The account plan will be developed by a combination of account team, product lines, senior management, and, at some stages, the customer.

Conceptually, the account development strategy matches the SWOT of the customer and of the telecommunications-based service provider.

As stated in Section 6.2.3, "If you can consider the different PESTs seen by suppliers, yourself, and your customer, you can also consider different SWOTs. The simple, yet powerful, insight is that the weaknesses of other members of your extended value chain provide opportunities for you to provide a solution."

Customer service, including the use of a customer service plan, is covered in Section 11.12. See also Section 7.6.2.

9.8 MAJOR BIDS

9.8.1 What Is a Major Bid?

A *major bid* is usually a transaction that has one or more of the following characteristics:

- *Large:* Size is relative to the scale of the service provider's total revenue. Ten percent would make a good starting point. One million dollars might also be a measuring point. A *mega bid* (as discussed later) would be one that provided a huge strategic opportunity. A large bid may also be one that will take a disproportionate amount of resources (constrained technical or other manpower, systems, support resources), thus requiring careful planning and staging.
- *Complex:* A bid is complex if it employs new technology, a new combination of technologies, or addresses a new/complicated application. It may be that new countries are to be supplied. Complexity can also come from the required terms and conditions: many major bids now require the supplier to become a risk partner. A complex bid also may be one that requires nonstandard support arrangements.
- *Strategically important:* There are some pieces of business that just *have* to be won for market positioning, defense of existing business, weakening or even annihilation of a competitor, or to gain a foothold in a new-name account.
- *Written down:* A major bid will almost certainly be the subject of a written statement of requirement from the customer—or a series of such documents. The supplier, for his part will surely provide a

written proposal. The absence of a written customer requirement would indicate a high, normally unacceptable level of risk. Customer expectations and supplier capability will be far apart; the costs of bringing them together will be high for both parties. The thorough establishment of requirements and solutions is the single most important factor in the successful practice of service provision—a subject handled in detail in Chapter 11. For now, see briefly Figure 10.3.

9.8.2 The Four Phases of the Major Bid

The golden rule of winning major bids is this: *If you don't know about an opportunity until you see the written request for tender, then you will almost certainly lose to someone who did.* You have to get in early. Figure 9.4 charts the process.

9.8.3 Initial Actions (Phase 1)

In carrying out predetermined marketing/business development activities, an opportunity is spotted. In particular, an opportunity is identified as a result of dialogue with potential and existing customers. Take every opportunity to understand what your customer needs. Assess these needs early; share them inside the business with senior management and product specialists to see how well the needs map on to the telecommunications-based service provider objectives and capability.

Share *white papers* with the customer as opportunities begin to develop. Analyze areas in question, raise issues, and propose solutions and approaches; provide product documentation, seminars, and demonstrations; reference site visits. Ensure contact is maintained at all necessary levels and with influencers, decision makers, end users, and budget providers. Watch the customer's timetables.

Start drafting the *bid prospectus*. The first draft will include an assessment of the opportunity and the likely solution. It will focus efforts on selling the benefits of the preferred solution—not the one that will not work or that may be the one preferred by a competitor. Identify the *hot buttons* of decision makers and influencers. Establish the critical success factors. Solution plus benefits plus critical success factors provide the basis of the *win strategy*. Consider pricing issues.

If at all possible, provide the customer with a skeleton, pro forma *invitation to tender* (ITT) document. This is a fundamental method of encouraging a focus upon the areas handled best by the telecommunications-based service provider and worst by competitors.

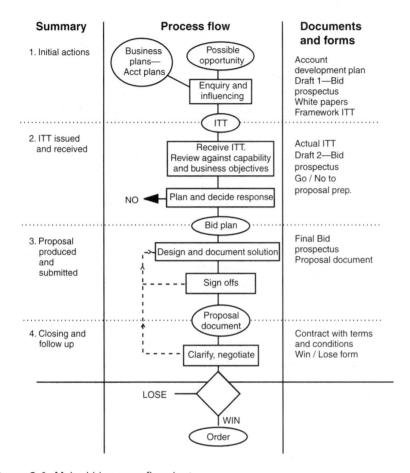

Summary	Process flow	Documents and forms
1. Initial actions	Business plans— Acct plans / Possible opportunity / Enquiry and influencing / ITT	Account development plan Draft 1—Bid prospectus White papers Framework ITT
2. ITT issued and received	Receive ITT. Review against capability and business objectives / NO ◄ Plan and decide response / Bid plan	Actual ITT Draft 2—Bid prospectus Go / No to proposal prep.
3. Proposal produced and submitted	Design and document solution / Sign offs / Proposal document	Final Bid prospectus Proposal document
4. Closing and follow up	Clarify, negotiate / LOSE / WIN / Order	Contract with terms and conditions Win / Lose form

Figure 9.4 Major bid process flowchart.

9.8.4 The ITT Is Issued and Received (Phase 2)

All being well, an ITT will be received. Waste no time in assessing it and deciding whether or not to bid. The bid strategy must be considered and documented in the next iteration of the bid prospectus, which is in effect a draft bid document, business case, and a refinement of the win strategy. The bid team is assigned and a timetable agreed. As well as the normal account team, there are three people with specific roles:

- *Bid manager:* Responsible for producing and submitting an approved, timely, and compliant bid that delivers the win strategy.

- *Technical design authority:* Defines and authorizes the technical solution. Also, similar roles exist for the implementation and for customer service.
- *Release authority:* Defined member of the management team.

9.8.5 The Proposal Is Produced and Submitted (Phase 3)

The proposal is produced to project winning "themes" that address the customers requirement and meet the critical success factors. The document will meet the telecommunications-based service provider's branding standards as defined by the marketing function. An author and reviewer is assigned for each section. Unless the customer specifies otherwise, a bid will normally contain the following sections:

1. Covering letter—signed by account manager but supplemented by senior management as required;
2. Management summary;
3. Summary of customer requirement;
4. Proposed solution;
5. Implementation and quality plan;
6. Operation/customer service;
7. Investment summary/pricing;
8. Terms and conditions;
9. Supplier profile—report and accounts/experience statements/reference customers;
10. Product brochures;
11. Bid production;
12. Other supporting literature.

Usually, there will a part of the ITT that describes highly specific requirements, both mandatory and desirable. This will require a section that has nothing but point-by-point direct responses, such as: *compliant, partially compliant,* and *compliant with development/when more details supplied,* with text to explain. Many bid documents have too much "sizzle" and not enough "sausage." One rule to bear in mind is that every sentence should contain either a fact, a name, or a number.

The document is *read by as many people as possible,* especially with detailed reference to customers requirement, and then it is released and delivered with time to spare.

Two tips on bid documents follow:

1. If, by mischance or bad planning, the submission date is going to be missed, then at least put in a summary with a price that will

keep you in the race and a commitment to provide further information within the framework and parameters of what has been submitted.

2. Also, submit additional small-size proposal documents—they will be the ones that get carried around in briefcases and read on trains and planes.

9.8.6 Closing and Follow-Up (Phase 4)

The submission of a bid is just a step along the way. The whole process of customer assessment, shortlisting, further discussion, capability demonstrations, and *proof of concept* must be undertaken. Terms and conditions and prices must be negotiated and agreed. Success rates vary. One in two or three is normally good. No win in five means it is time to think again. The hungrier the supplier gets, the more chance they have of winning. The more business a supplier gets, the more its capability may become stretched.

Success at each point should be measured. Reasons for winning and losing should be documented and fed back into the product, marketing, and sales processes.

Finally, honor the bid team and tell the world about the wins.

9.8.7 Major Bids—Summary

The major bid is only a part of the overall success in managing a large service provider customer. There are a number of repeatable principles that, if applied, will do much to ensure a high success rate:

- Spot the opportunity early.
- Qualify technically—"Can we do it?"
- Qualify strategically—"Why is this important, even vital, to us...at what price?"
- Qualify from resource and *opportunity cost* standpoint—"What will it take to do this properly?" "What will we not be able to do if we do this bid?"
- Train people to write winning proposals.
- Get a good format, to a high production standard.
- Build the team early, with a strong bid manager. Cosset the team: It will be working long hours under pressure.
- Follow the process, use pre-prepared material but remember what this customer is asking for, since every customer and bid is different from the last.

9.9 DISTRIBUTORS

Distributors are like the telecommunications-based service provider's own sales people or remote branches. They have special needs:

- They are probably located at a distance.
- They will generally appreciate a regular contact by telephone—backed up by the efficient management of such issues as brochure supply, new software, training course bookings, end-customer visits to the home-base of the service provider, and so on.

Many organizations send senior managers and sales people to visit, but forget these simple tasks. But distributors may be an essential ingredient in a multichannel contact strategy. They must be kept up to the required standard and supported accordingly.

Service providers are by definition distributors of the technical and information elements provided by their suppliers. It is as well to insist on the same treatment from those suppliers as you are planning to provide to your own agents and distributors.

A key "distributor" in the telecommunications-based age is the browser or search engine of the Internet.

9.10 SYSTEMS INTEGRATORS

The purchase of the skills and resources of a systems integrator in developing telecommunications-based service provision products has already been described in Chapter 8, but there is also the possibility that the service provider will have a sales relationship with a systems integrator. Special care is needed in managing such a relationship.

Systems integrators may act as a channel to market for a telecommunications-based service provision, since they may integrate its offering into an overall information management solution for a large client, such as a bank, an airline booking function, or a major call center with a requirement for links to information sources right across the Internet. Given the increasing tendency for organizations to partner with each other, the increasing scale of information systems integration activity and the close relationships that systems integrators forge with their large corporate clients, it is important for a telecommunications-based service provider to know how to manage the relationship with them.

As well as linking with each other in both purchasing and selling relationships, systems integrators and service providers are akin to one another in many ways. They carry out a similar functions, but they are distinct from each other. Service providers operate the value chain introduced in Chapter 4 and developed throughout the book.

For example, referring back to Section 4.10, both telecommunications-based service providers and systems integrators normally do the following:

- Create services that comprise information, technology, and other elements and present them to customers in an easily used manner.
- Act on behalf of their customers in selecting the suppliers of the elements and act in alliance with their suppliers, while remaining free to change them if it is beneficial to do so.

Systems integrators operate the systems development and deployment methodology shown in Figure 9.5. The terms are self-explanatory.

Systems integrators, however, are normally associated with large, complex projects for corporate customers. They generally, but not always, stop short of building and deploying packaged products to defined markets. (If they do so, they normally form a discrete business focused upon that task—as a service provider.) Systems integrators normally compete for business in a short-timescale competitive-bid situation, as described in Section 9.8. Following months or even years of developing a relationship with the potential customers, a major bid implies a major investment of time and emotion into a single chance to win a large piece of business that represents a significant proportion of the year's target. On the other hand, the typical telecommunications-based service provider has many opportunities to win smaller pieces of business.

Systems integrators, like internal major bid teams, often appear abrasive and unreasonable to their suppliers. Here are some typical needs and problems:

- The systems integrator's people will be unfamiliar.
- There will be demands for statements on functionality, availability, and price. These may well be the same as the demands the telecommunications-based service provider, itself, brings to its own suppliers, but they will be couched in unfamiliar terms or cover rarely covered areas.
- Time will be short; resources will be limited. Bid teams become nervous and impatient.
- The systems integrator's client will probably require performance guarantees. Extreme care must be taken to ensure that the company itself is not put at risk by agreeing hastily to barely understood terms and conditions.

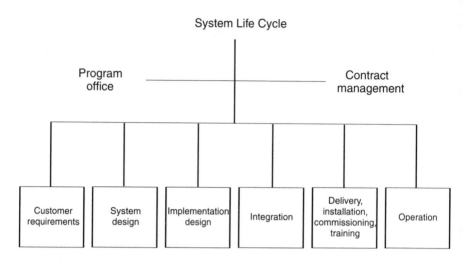

Figure 9.5 Work breakdown structure—systems development life cycle.

- Language and technical jargon and project management methods and practices may all be unfamiliar. Information systems will not talk to each other.

To make a success of selling through a systems integrator you should do the following:

- Appoint a relationship manager with the skills to anticipate and translate bidding requirements into your own language.
- Establish electronic links, application software compatibility, and document standards early on.
- Create parallel peer relationships as early as possible (e.g., introduce contract managers to each other).
- Be clear on what it would take to simply walk away from the deal and relationship.
- Follow all of the other rules already given for major bids (e.g., early warning, pre-prepared material, resource allocation, team building, frequent management briefings).

9.11 GETTING THE OUTPUTS RIGHT

The sales process outputs are the implementation process inputs. Sales is a process that mixes *fuzzy* and *rectangular* activities, traditionally with the emphasis on the former. The implementation process is 95% rectangular.

A salesperson who is 51% successful in converting possibilities into sales is likely to be deemed highly effective. Sales people are measured primarily upon their achievement of volume and margin.

Implementation people have to put solutions together that are 100% right. An error of one digit in the customer address will result in no implementation at all; an error of one digit in the delivery date will create an angry customer and at least a doubling of delivery cost. A fault in configuration of an information-based product will cause it not to work. Errors in the input of pricing data will lose money forever in the relationship. Overzealous promises regarding support available will result in additional costs—or a shorter customer relationship—so less overall return. Finally, the arrival date and time of sales orders may be difficult to predict.

All sales ultimately happen because of a decision of by a customer, not of the telecommunications-based service provider. Call center resourcing is one of the new specialist skills of the telecommunications-based age. The number of *volume sales* of established products is normally predictable within limits, or at least subject to known factors and influences. Larger customers are fewer in number, so there is less *portfolio effect* to aid predictability. Sales to larger customers are more "lumpy."

There is constant tension between sales and implementation. It is inevitable, but it must be managed and channeled into good.

The solutions to these problems are contained within the statement of the problems themselves. This is what must be achieved:

- The right solution for the customer—properly configured.
- Excellence in documentation.
- Early warning of what is coming—based upon promotional activity, time of year, good large customer sales forecasting, and so on.
- Early involvement of members of the implementation team. If they can be there to frame the solutions and to coach sales people, it will help.

Should the implementation manager be given power of veto on orders? Certainly there should be a sign-off as part of the sales order process, but the key is in early involvement in a *positive* manner.

9.12 MEASUREMENT AND RECOGNITION

The process outline listed possible ways to measure. They require little further comment:

1. Opportunities identified;
2. Times "short-listed;"
3. Orders achieved;
4. Ratios of 1, 2, 3;
5. Gross sales volumes/value;
6. Gross margin (initial order and overall revenue);
7. Terms and conditions achieved;
8. Length of contract;
9. Accuracy and quality of documentation;
10. Accuracy of forecasting;
11. Cost of sale.

The acid test is the present value of gross margin achieved.
The important principles to follow are as follows:

- Align sales team rewards with what is profitable for the company.
- Remember that recognition has a key role to play.
- Honor all, not just the sales teams.
- Reward accuracy—not just volume. Inaccurate documentation reduces margin.
- Remember that 90% accuracy and tidiness with $10 million sales will always beat 100% accuracy with $5 million of sales.

9.13 RISK, COST MANAGEMENT, AND DISASTER RECOVERY

The greatest risk *in the sales process* for the telecommunications-based service provider is that there will be too few sales, or no sales at all.

The next risk is that the sales obtained will be for the wrong products at the wrong price in the wrong place with the wrong service arrangement.

Loss of reputation, product liability, and serious financial damages constitute the next risk.

A U.S. boss of mine described it to me as "Some damages etc. are a silver bullet through the heart of the corporation; some chop off its arms and legs so that it may die anyway..."

Finally, there is the risk that the sales obtained will be documented and configured poorly and so transferred into the implementation process as a nightmare customer relationship to come.

These risks can be contained only by the application of detailed process, coaching and supervision, and clear lines of authority. In a telecommunications-based service provider, the customer-facing decisions usually have to be made fast.

Costs can be controlled best by ensuring focus upon the right marketplace by the effective design and management of channels, and the incorporation of sales costs into the incentive scheme. As mentioned above, many help people and teams concentrate upon that, but most of the costs are "hard-wired" into the sales process with little left as discretionary spending.

There are some disasters that can take place: loss of key people (including those who might go to a competitor), loss of phone lines to the call center, loss of functionality and information from crashed systems, and so on. All of these can be planned for and accommodated, as detailed throughout the book.

But, despite all, the worst disaster of all remains no sales at all.

9.14 CONCLUSIONS

This chapter has covered a range of subjects. First, there has been an intention to sketch out the sales process for those for whom it is unfamiliar. Second, standard sales practices has been related very specifically to telecommunications-based service provision. Third, the chapter has shown how sales to large customers and volume customers may have the same underlying process, but that the execution through different channels is vastly different. The processes built upon both call centers and upon the major bid process have been described in detail. All of these factors must be considered within the product concept from the start.

It is clear that a well-executed sale is a first step to a well-implemented solution and a profitable, satisfied customer. Failure to follow the process will inevitably lead to confusion and pain. The telecommunications-based service provider, however, has all of the information and on-site telecommunications and computing to give sales people the power to win.

A good understanding of what is involved is essential for a poorly executed sales process will lead inevitably to a poor installation/delivery of the product and an unhappy customer who will not want to pay the bill. It is generally prudent to include installation and service people in the process of approving the sales about to be made; if they are not included, then they will be reluctant to own the situations they inherit.

Sales forces often sit uneasily with the rest of the company. Some people hold a "folksy" view that sales people are extroverted, meretricious, overpaid, self-centered people of doubtful ethics and cavalier attitudes to thoroughness or reality in execution. We all know people like that—but they are no more prevalent in sales than in any other area that requires a juggling of expectations and resources. Good sales people really earn the salaries and other benefits they command in the market-place. In a badly run company, a sales team thinks it is "kicking custard whenever it tries to get something done." Sales people want to belong to a winning team.

Sales people are used to working away from the rest of the team, with little information at their disposal, but we can now bring them into the overall team better than before. To their flair we can now add excellence in information management and telephone-based support.

Implementing Orders **10**

10.1 OVERVIEW

The implementation of orders, once placed, may produce the first tangible manifestation of what is being sold; it will certainly be very specific to the purchaser and the satisfaction of her requirement. This is the stage where the transaction moves from being a possibility and becomes an integral part of whatever it is that constitutes the competitive proposition for a business user or lifestyle solution for end consumer. The feelings, motivations, and actions of the customer are driven accordingly.

Order implementation is the vital link, as shown by its position in the overall service process provision process in Figure 10.1.

10.2 PREMIERE WORLDLINK AND PAY-PER-VIEW

Having decided to buy a product or service, most of us want it straight away. The new telecommunications age abounds with examples of how this is now being achieved.

Case Study: Premiere Worldlink

Flicking through the United Airlines in-flight magazine as I crossed the Atlantic, I found an advertisement for a "Premier Worldlink" phone card associated with a *call-back* service that would provide, immediately, the benefit of competitive U.S. tariffs in 50 or more countries of the world. Why my interest at that moment? Still bruised from a $1,000 telephone bill in an Eastern European hotel, I was motivated to find ways of achieving a more favorable rate for my calls away from home, starting with this trip. How was the service activated immediately? A credit-card sized,

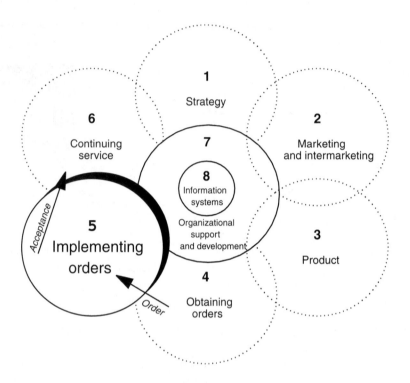

Figure 10.1 Process area 5—implementing orders.

knock-out section of the advertisement acted as a start-up card. I telephoned the Worldlink call center as soon as I arrived in the United States (a toll-free 1-800 number) and supplied credit card details. Thus, issues of identity, credit worthiness, and billing were handled at a stroke. I was offered a choice of account numbers and chose one based upon other familiar numbers of mine. With the addition of a PIN and some private personal information, I was ready to start calling.

You can buy a prepaid phone card in a shop and use it immediately; similarly, you can use one of the routinely accepted credit or charge cards to pay for a call in a suitably equipped payphone. What distinguished the presentation of the Worldlink product was that it also created a customer relationship that could endure and a database record that would provide the means of extracting added value from up-selling and cross-selling as the occasion arose.

If the technology of switching and billing credit card phone calls is now well established, the arrangements for buying and viewing a single pay-per-view television program are still relatively new.

Imagine, a few years ago, calling the telephone operator and asking for a movie or other television program to be transported to you at broadcast quality. Given that we now have the technology to store and transport such pieces over ever-decreasing mediums of bandwidth, we still need a cost-effective process to select, activate, and pay for them. The product is short-lived and the margins are slim. There is certainly insufficient fat in the price-cost equation to do this manually. What was needed to make that possible was a real-time or near real-time automated process, with the labor content being provided by the customer, not the supplier.

Case Study

It *is* done by one cable TV company in North America, using Periphonics interactive transaction processing. Note, incidentally, that the interaction can be carried out with a telephone, using dial tone multifrequency tones(DTMF) or voice recognition or by using a computer terminal linked to a WWW home page. The *tree and branch* IVR application developed for the first can be used directly for the second, too.

10.3 PROCESS DESCRIPTION

The implementation process converts a customer order into a working product or service that is available to and usable by the customer and that can be billed and supported by the telecommunications-based service provider. The activities to be carried out are shown in the process description table—Table 10.1—and detailed throughout the chapter.

The three main issues of implementation are those of timeliness, accountability, and accuracy in meeting requirements and cost. There are particular distinctions relating to all of these in the context of telecommunications-based service provision.

If the process of implementation is carried out well, then the telecommunications-based service provider will have a properly equipped, well-trained customer who is motivated to make good use of his product or service and who will be predisposed to remain a customer over a long term, perhaps increasing his engagement by trading up and/or across the product range.

If the implementation process is carried out badly then the result will be one or more of customer dissatisfaction, extra cost/reduced profit, order cancellation, and damaged reputation.

The implementation of telecommunications-based service provider products instantaneously to a volume market and the long time cycle delivery of a complex project to a major customer operate to the same basic principles, though the actual execution varies enormously across that continuum. This is illustrated in Figure 10.3.

Success in implementation comes when:

- Delivery proceeds to the agreed timescale.
- All changes to requirement or design are communicated and documented effectively.
- The product does what the customer wants it to do, and more.
- Customer service records are created and checked.
- The telecommunications-based service provider creates accurate billing records and acts upon them.
- The service provider ensures it does not agree to any arrangement or contractual commitment that will harm the organization.
- The customer signs an acceptance certificate and pays a bill to that effect.

The formal process is described in Table 10.1 in the same systematic form as earlier chapters:

Table 10.1
Process Outline—Implementing Orders

Field	*Attribute*
Process description	Implement orders
Process owner	Implementation of project management using resources for application development, manufacturing, integration, installation, customer assistance, and training

Field	Attribute
Subprocesses and activities	Quote from Chapter 10: "It is essential that issues of order implementation and customer service are identified and handled prior to an order being accepted, or a nonstandard product, price, or condition offered."
	The following terminology relates to a major project. The equivalent activities for an instantaneous order implementation are exemplified in Figure 10. 3.
	Kickoff meeting Procure required information or technology elements to build service Assign people to build and deliver service
Time to complete	"Instant/on demand" "Urgently/as soon as possible" "On date X at time Y" "90 days" "No hurry"
Inputs (and sources)	Sales orders ("obtain orders" process) Product elements (various sources)
Outputs (and destinations)	A completed installation, accepted by customer, with billing and service information captured ("provide continuing service" process)
Lead process area for these information systems...	Implementation project management Logistics systems—stores, delivery
Support documents	Implementation procedure document Project statement of works Detailed customer information Solution design specification Project plan—including GANTT charts, task lists, dependency diagrams, etc. Order and delivery documentation for information and technology elements supplied to the telecommunications-based service provider. Change control documentation Time sheets, expense claims Training documents, help screens, and voice prompt documentation Customer acceptance certificate Handover notice from implementation project team to customer service team Bill to customer
Effectiveness measurement (measured by those receiving the outputs)	Budgeted and actual cost Contracted and actual elapsed time Backlog Customer satisfaction

Table 10.1 (Continued)

Field	Attribute
Risk and cost management/ disaster recovery	Risk and cost management: Liquidated damages Contingent liabilities Differential terms with customers and suppliers. Need to ensure "back-to-back" arrangements to cover nondelivery of components, excess price movements, product performance, etc. Stockholding risks and costs Costs resulting from poor planning N. B. The following issues were identified in Chapter 9: emphasis on prevention not cure—especially on pricing, deliverables, terms, and conditions Disaster Recovery: Ensure that information on current implementation projects is protected and recoverable—the orders already received are those that will provide revenue first.

10.4 TIMELINESS—CHANGES IN EXPECTATIONS

There are four sorts of time dimension for consideration by the telecommunications-based service provider : "now," "as soon as possible," "on date X at time Y," and "no hurry." The way these interact with the issues of implementation/delivery and the way they interact together are discussed below.

10.4.1 A Change From Producer-Driven to Market-Driven

A traditional standard of the elapsed time from order to installation for the many products and services to be installed has been, say, "90 days" or "subject to product availability." Delivery may have been offered only for "9 to 5, weekdays only." Configuration and tailoring of products to individual tastes or requirements was a rarity. Hallowed industry practices, developed in many cases for the supplier's own benefit, tended to be accepted by customers.

This has largely changed in the age of the telecommunications-based service provider. Technology advances, competition, and increasing customer sophistication and assertiveness have driven down waiting times and reduced tolerance of inconvenient supplier working procedures.

Time is clearly of the essence in the case of a service such as pay-per view. Delivery of many telecommunications-based services may be near instantaneous, as in the "Premiere" service described above, or in any telephone call for that matter. But quick delivery is increasingly crucial for products in general. Once having decided to buy, most customers want their product or service without delay, delivered at a time that *they* specify. What is more, customers expect their products to do more for them; they expect them to be delivered tailored/preconfigured without any impact to the delivery time profile.

10.4.2 Tradeoffs—Cost Versus Time/Convenience

There are often fundamental tradeoffs between speed of delivery and cost.

Case Study

The leisure traveler on a competitively priced package holiday may book well in advance for a set date and then travel on a 3 a.m. flight from a second-level airport, arriving two hours before takeoff. Then, at the destination airport, will wait in a half-full coach for the whole party to gather, even waiting for part of the party to arrive in another aircraft from another departure airport. The coach, now full, will drop off sections of the party at each of the several resort hotels used by the holiday company before, at last, the leisure traveler is dropped at his own. Similarly, the student on an adventure trip to India may be willing to risk missing a flight in return for the substantial discount associated with a *standby* or *advance purchase* ticket.

The business traveler can rarely accept any of those regimes. Arriving at a convenient airport by taxi from home or work, the business traveler may even walk straight up and buy a ticket just ten minutes before a flight with preguaranteed capacity for such an eventuality. The flight is timed to meet her business-hours requirement. She takes a taxi as soon as she arrives at the far end. The business user pays considerably more for the flight than the leisure traveler.

The airline is able to maximize its efficiency by blending these diverse requirements into its scheduling plan. Postal and courier services adopt similar "class of service" approaches. The example illustrates each of the instances of "now," "as soon as possible," "on date X at time Y," and "no hurry."

In prioritizing countries for accommodation in global markets, either as producers or as consumers, the efficiency and convenience of physical transport links is a key determinant. So are telecommunications and the sophistication of telecommunications-based services. Staying with the example of the business traveler: a journey can be booked and paid for in a 5- to 10-minute phone conversation with a highly trained agent online to an (or to several) airline computer system(s). But this cannot be done in the large percentage of the world, where you have to go downtown with cash, wait in line to talk to a travel agent with out of date schedules and price lists, and then wait a day or two before you discover whether you have a seat. The overall "order" and "delivery/implementation" elapsed time varies enormously between these extremes. Section 5.4 discusses the concept of *fast cycle.* Refer right back, also, to the start of the book where a telephone call was made to a number of airlines to test their relative effectiveness.

10.4.3 Delivery at a Specific Time

Business users of telecommunications say that it doesn't matter so much *how long* it takes for, say, a private long distance digital circuit to be installed, so long as it is *actually delivered* on or by the date promised, so as to fit with overall project schedules. A private user who books half a day off work and is unpaid to be at home for a delivery is understandably irate when the promised delivery does not come.

This issue is self-evident. The solution lies partly in good communication, discussed in Section 10.5.

10.4.4 No Hurry

The leisure traveler will probably enjoy strolling to a high-street shop to mull over possibilities and select a holiday destination; the coach ride around the resort may represent a good chance to orient oneself and spot restaurants or beaches for later exploration. The globe-hopping student may appreciate the opportunity to sit at an airport and catch up with post-cards, to meet fellow travelers, to learn how an airport lives. While few business customers would see things in that way, it is wrong to conclude that *business* must always mean fast.

A colleague of mine comments that people seem to expect instant replies to faxes, whereas a lapse of a few days to answer a letter is deemed acceptable. Instant and ubiquitous personal telephony can exert a tyranny of overcommunication upon personal and business relationships. There is a phenomenon known as *dueling at dawn,* whereby two or more people ping-pong e-mails at each other. Voice mail can be the same. Instantane-ousness is a new opportunity for the "workaholic."

People need time to think, and in many circumstances they actually want to take more time over what they are doing ("to smell the flowers along the way").

10.5 TIMELINESS IN DELIVERY—INCREASING CAPABILITY

But on the whole, people appreciate the opportunity to get things done without delay; there is a capability to do so.

Advances in technology and a greater willingness to carry out trans-actions over the telephone or by e-mail have contributed to increased ca-pability. There is an virtuous spiral of timeliness caused by the existence of more and more arrangements that can serve as the basis for the expe-dited provision of something new. In the Worldlink Card example (in Section 10.1), it was the preexistence of a valid credit card account that made immediate activation possible.

Even where there are physical elements to be delivered, too, it is possible to construct a clever combination of process to achieve near-in-stant delivery.

Case Study: Ameritech

If you walked into a mobile phone dealer in the Chicago area as far back as 1992, you could select the handset and walk out with it working on Ameritech's network. The network service startup requires credit check-ing, tariff choice, and activation at the switch for the selected instru-ment's electronic and physical identity. The credit checking questions are asked first; the Ameritech call center's subscriber management system goes into a parallel/background mode, dials the credit agency, and inter-rogates its database; nine times out of ten the credit "clear" is received before the other actions have even been completed. Similar methods will be found in most advanced telecommunications countries now.

Telephone networks are capital-intensive. In earlier times, when-ever a subscriber requested service termination, the associated line plant and subscriber equipment were disconnected and reused elsewhere. Over time, capital costs went down and labor costs went up and it became more economical to leave equipment in place, still disconnecting service at the main distribution frame in a central office (public telephone exchange). Later still, the service could be activated/deactivated at the switch. This gave the opportunity for very much quicker supply of service to new customers moving in.

Case Study: A European PTT

(*Nota bene:* The subject of hub-and-spoke information systems and online transaction processing is covered in more detail in Chapter 13 (see Figure 10.2).)

All of these factors were as true for a particular European PTT, but there was a problem. If the telephony systems had come up to date, the network management and subscriber management systems had not. They operated in a range of unconnected and often proprietary computer- and paper-based environments. The *network management systems* (NMS) and *subscriber management systems* (SMS) all held their own data and operated applications for different purposes. They were not adequately documented. The systems were not linked—or rather, nobody knew whether or how they might be linked. Somewhere deep inside the telco the information was being used for other purposes (such as credit control or international settlements). It was just not possible to switch these systems off. In any case, they worked and they represented a massive asset of data

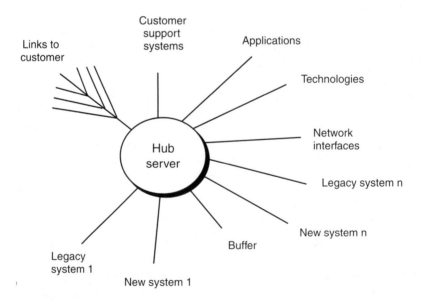

Figure 10.2 Hub and spoke diagram.

already input. As a result of the generally leisurely approach to providing service to new (or moving) subscribers, the individual subprocesses mounted on these systems were slow and often were carried out in sequence. Notwithstanding the advances in telephony, it could still take 20 to 30 days to provide service. Rigorous re-engineering of each process brought the time down to 10 days.

But the PTT now provides service in two days and often less than one. How? Each of the legacy systems was connected to an *online transaction processor* (OLTP), which acted as a central server. With a fair degree of data and systems re-engineering, using the automated outputs of one process to act as the inputs for another, it became possible to get data buzzing around fast enough for the formerly end-on-end systems to support parallel processes at near real time.

10.6 ACCOUNTABILITY

In the essence of organizational success in telecommunications-based service provision is the requirement that there should be clear accountability for each part of the overall process. Deeper consideration of the order fulfillment process gives a further perspective.

10.6.1 Clear Description of Processes, Inputs, and Outputs

The essence of a team approach for a telecommunications-based service provider is that the *implement orders* team relies upon the *obtain orders* team to ensure the following:

- The customer's requirement is understood.
- There is an agreed definition of how those requirements will be met.
- There is commitment from all those inside and outside the organization to supply the goods and services needed to provide the solution, and an order is placed to that effect.

In Section 10.3, six criteria for success for implementation were noted. Ipso facto, these become the "accountabilities" of the "implementation manager." They are listed again below:

- Delivery proceeds to the agreed timescale.
- All changes to requirement or design are communicated and documented effectively.
- The product does what the customer wants it to do, and more.

- Customer service records are created and checked.
- The telecommunications-based service provider creates accurate billing records and acts upon them.
- The service provider agrees to no arrangement or contractual commitment that will harm the organization.
- The customer signs an acceptance certificate and pays a bill to that effect.

The above is so that the "provide continuing service" team can ensure the following:

- Both customer and service provider meet their financial objectives.
- The customer is motivated to be in a long-term relationship.

10.6.2 People Often Belong to Several Process Teams

People are sometimes "owners" and sometimes "contributors." Are the teams entirely discrete? Does one set of people in the "implement orders" (us) team wait mindlessly for other people in the "obtain orders" (them) team to deliver a perfect output at precisely the forecasted time? Can the "obtain orders" people blindly demand perfect delivery from the "implement orders" team?

Sales and delivery teams *do* fight (in Chapter 9, see Sections 9.11 and 9.14).

But the organizational design described in this book recognizes that people need to belong to several of those teams—having "process ownership" in some cases or being "contributors" in others. Hierarchical rank is a secondary consideration. Knowledge, information, experience, and opportunity to help are the key drivers to the dynamics of such a team.

As an example, consider a senior, highly experienced technical individual (M.S., chartered electronic engineer) with 15 years business experience, with former experience of working in banking, with specialist knowledge of channel *associated signaling* (AS), interested in yacht cruising, who could simultaneously occupy the roles shown in Table 10.2.

From the discussion and the table above, it will be seen that, for example, *sales* will have a continuing role to play in the Z Corporation Project after the order is placed—contributing to communication with the customer, managing expectations, and being part of dialogue on changes. The project manager will surely have contributed earlier in the cycle, during the *obtain order* process activities to understand the customer requirement, to define the solution, and to ensure that it would be deliverable should the order be placed. The project manager will have had every

Table 10.2
Titles, Roles and Process Team Membership

Title	Role
Implementation project manager	Accountable for Z Corporation—theater seat booking system
Member, business strategy team	Contributor bringing general technical insight ("strategy")
Industry specialist member, financial services business team	*Contributor* of industry-specific experience, knowledge, and contacts ("marketing/intermarketing,," "obtain orders")
Technical specialist, AZ telco project	*Contributor* of specialist CAS knowledge to team seeking to win "AZ telco project to deliver IVR as part of intelligent networking service creation" ("obtain orders" and, if an order is placed, "implement orders")
Yacht skipper, XX telecommunications-based service provider corp. cruising club	Contributor of an interest and commitment to the corporation-sponsored club for its employees Owner, when actually acting as skipper on a boat of the club ("support the organization")

opportunity (and indeed a duty) to remain informed on the likelihood and timing of an order.

The role of implementation project manager is pivotal to the reputation of the firm; it is indeed a tough one.

10.7 ACCURACY AND COMMUNICATION

The accurate progression through the value chain from the end of the obtain order process to the start of "provide continuing service" is fundamentally dependent upon communication.

The intrinsic purpose of the sales process is to take initially unclear customer needs and expectations and associate them with the most suitable of the telecommunications-based service provider's available products. The process described in Chapter 9 comprises preparation, inquiry, presentation, close, and administration. These stages and subprocesses are all present whether the transaction is long or short, large or small. The obtain order process is *fuzzy* at its start when needs and solutions are known, respectively, to only one of the two parties. The character of the process becomes increasingly *rectangular* as information is exchanged and the process moves to its end.

By the time the order is signed, both customer and telecommunications-based service provider should have a common view of what the former is going to get, what it will do for him, when he is going to get it and how much it will cost. In practice, there is a large area of uncertainty. This is illustrated humorously by Figure 10.3.

The task of converting a customer order into an accurately working product or service clearly requires a high degree of communication with the customer to manage expectations to a mutually acceptable point.

The implementation of each customer order is a project—a series of linked tasks to achieve a stated aim within the required budget and timescale. The communication of project progress, proposed changes, and so forth is an integral part of project management. Communication requires listening skills and willingness to compromise.

Communication is not just between project manager and customer, but also along the chain of people from salesperson to storekeeper to delivery driver to billing clerk. In more complex cases, the chain includes system designer, manufacturer/configurer, trainer, and so on.

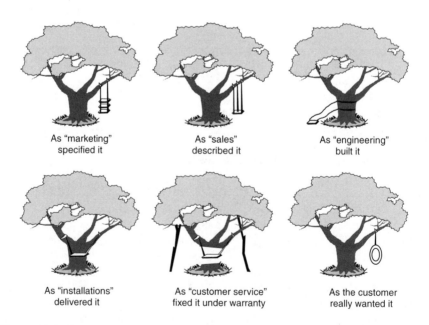

As "marketing" specified it

As "sales" described it

As "engineering" built it

As "installations" delivered it

As "customer service" fixed it under warranty

As the customer really wanted it

Figure 10.3 Customer expectations.

Communication needs common language, media, and documentation. Indeed, the following examples of documentation define the very process of implementation, not just the element of its communication:

- Price and product lists and details;
- Customer requirement statement;
- Configuration notes;
- Order form—duly completed and signed;
- Implementation procedure document;
- Project statement of works;
- Detailed customer information;
- Solution design specification;
- Project plan—including GANTT charts, task lists, dependency diagrams, and so forth;
- Order and delivery documentation for information and technology elements supplied to the telecommunications-based service provider;
- Change control documentation;
- Time sheets, expense claims;
- Customer acceptance certificate;
- Handover notice from implementation project team to customer service team;
- Bill to customer.

The documents of communication and record define the process itself. A method of putting together the process and its constituent activities in practice is shown in Section 10.10.

Throughout the implementation process, it must be remembered that the customer needs to be continuously reassured that he or she has made a good choice.

10.8 GETTING IT RIGHT—COMMERCIAL AND LEGAL PERSPECTIVES

There are many things that can go wrong in converting a customer order into a working product or service that is available to and usable by the customer and that can be billed and supported. If there is a well-defined customer requirement, a properly functioning product that meets the requirement, and a constant flow of information between service provider and customer with regard to such issues as delivery date and cost, then all will be well.

The commercial and legal perspectives upon implementation are as follows.

1. There should be a clear and common record of what is to be achieved, when, and at what cost.
2. There should be a systematic approach to putting things right if they do go wrong.

Commercial and legal specialists should advise upon the procedures and documentation relating to the former, since it will be they who reap the whirlwind and manage things when they go wrong.

Essentially, the supplier has a duty to ensure that the customer's requirement, as stated, will be met by the product being offered. To what extent this is achievable depends on both parties. Where the requirement is vague (either in fact or in the way that it is articulated) or where it can be seen that the product will fall short of meeting what it is clearly required, then this should be made clear. The precise legal position of the supplier in this regard will vary from country to country.

From the commercial standpoint, the telecommunications-based service provider will want to ensure that everything that is supplied is paid for; in the pursuit of "good service" many suppliers fall into the trap of providing things that are *outside* the contract and fall into jeopardy since they may not be paid for them and they may impact time and performance of those aspects that *are* in the contract.

10.9   IMPLEMENTING A MAJOR PROJECT

What defines whether a project is major? It might be so defined for one or more of several reasons. Building on Chapter 9, a major implementation project may be defined as containing the following:

- Large—perhaps greater than $500,000, or amounting to more than 5% of the year's expected new orders;
- New or complex technology;
- Complex contractual terms;
- Multiple sites;
- Requiring a *project manager* under the terms of the customer contract.

A major bid is almost certain to require a major project approach to deliver what is ordered.

A major project might also be a major upgrade exercise, either for one customer or to roll out a complete new release of software. Make a note that the arrival of the next millennium, 2000, will unearth all manner of unwelcome defects in software.

Case Study: 1996

Some computer programs show years as a two-figure notation. A woman born in 1892 recently received free education vouchers in a scheme to encourage four-year olds to start at nursery school.

Future Case Study: 2000 time 0000 and 1 microsecond

What has happened to the world? Whole telephone networks have come to a stop as their call rating software fails. Financial systems have just added billions of dollars of supposed interest arrears (from the year "1900") to savings accounts and bankrupted the banks who run them. Automated train timetabling and scheduling systems have canceled all of the trains for the following morning...

In any case, a major project is likely to be any order implementation (or upgrade exercise) that requires skills, knowledge, processes, or resources beyond those normally available to those responsible for standard implementations or customer service problems.

Major projects have a specially assigned manager to run them. The role of the project manager and the activities to be carried out are those defined in Section 10.9. The accountabilities and behaviors are those indicated by Section 10.4.

The telecommunications-based service provider is well advised to write a detailed major project procedure and to ensure without doubt that issues of scope and objectives , roles and responsibilities, authority, and resources are addressed fully in each case.

10.10 BUILDING THE IMPLEMENTATION PROCESS IN EACH CASE

It remains only to show how the content of the chapter can be drawn together to create an implementation process that will work for a particular telecommunications-based service provider and for a particular product.

The scope and objective of the order implementation process is to convert a customer order into a working product or service that is available to and usable by the customer and that can be billed and supported by the telecommunications-based service provider.

The process in each case must give full attention to the key issues of:

- Timeliness;
- Accountability;
- Accuracy in meeting requirements;
- Cost.

The table below indicates the actions to be carried out, its ownership, the associated documentation, the equivalent task with respect to a volume sale/instantaneous delivery.

As examples, the "Action" relates to a major project to deliver a financial information service (from a number of home produced and bought in databases) with the required hardware and desktop management, training etc. to make it work for a 500 user, 10 site customer. The *instantaneous service* is a telephone call made in association with a postpaid card

Table 10.3
Implementing Orders in Practice—Large and Small

Action	Owner	Documentation	Equivalent Action for Instantaneous Product
See row in Table 10.1 for subprocesses and activities		*See row in Table 10.1 for supporting documentation*	*A telephone call made in association with a postpaid card*
Write general "implement order" procedure	Written by implementation manager, agreed by CEO	Implementation procedure document, as part of the product specification	Write code to define how each call is to be handled when set up (ordered by dialing) by the customer, including rules for referral to help prompts or customer service agent

Action	Owner	Documentation	Equivalent Action for Instantaneous Product
Kick-off meeting to receive customer requirement and assess details of particular solution to be developed	Implementation project manager	Project statement of works Detailed customer information Solution design specific project plan—including GANTT charts, task lists, dependency diagrams, etc.	Receive information input (dialed or spoken) by customer Interpret it via route table for connection to distant end Check customer's account number, record, and PIN, and permit call to take place
Procure required information or technology elements to build service	Implementation project manager	Order and delivery documentation for information and technology elements supplied to the telecommunications-based service provider	Pass traffic over own circuits and plant and over those of partner telcos, in accordance with prearranged routing plans, interconnect agreements, and signaling system protocols
Assign people to build and deliver service, training, etc.	Implementation project manager	Project plan—including GANTT charts, task lists, dependency diagrams, etc.	If required, provide help in connecting call or in using system features
Advise customer of progress and expected delivery	Implementation project manager (sales, if required)	Progress reports—written and verbal—either sent to customer or made available as status reports on an accessible information system	"Your call is being connected
Offer alternatives or enhancements if required or appropriate	Commercial (sales, if required)	Change control documentation Invoices from subcontract/element suppliers	"I am not here at present

Table 10.3 (Continued)

Action	Owner	Documentation	Equivalent Action for Instantaneous Product
Assemble and integrate elements of the product	Application Development Manufacturing Integration	As required	Preset
Deliver to the customer		Delivery note	Call detail record created (CDR)
Train the customer	Training	Training documents, help screens, and voice prompt documentation	Instruction card, voice prompts, or via referral to customer service agent
Obtain customer acceptance	Implementation project manager	Customer acceptance certificate	Nonacceptance by exception only
Record use of resources, during the implementation, for purposes of billing and costing	Implementation project manager captures data Accounts record data	Time sheets, expense claims	CDR used to rate the call and create billing record Billing record referred across to tariffing agreement with customer (e.g., "is this a family & friends call charged at a reduced rate?)
Set up customer service and billing record	Customer Service Billing	Handover notice with all above order, contract, and project documents attached	Done at time of setting up service
Bill the customer	Billing	Bill	Bill
Hand over to customer service team	Implementation project management to customer service	Hand over acceptance notice	Call record available to customer service team as required

10.11 CONCLUSIONS

"High Technology obeys the Iron Law of Revolution; the more you change, the more you have to change. You have to accept the fact that in this game the rules keep changing" (Bill Joy, co-founder of Sun Microsystems, *Upside Magazine*, 1991).

Implementation of orders requires that iron law be tamed for long enough for the solution to be put in place and paid for, meeting the criteria of success below.

Success in implementation comes when the following occur:

- Delivery proceeds to the agreed timescale.
- All changes to requirement or design are communicated and documented effectively.
- The product does what the customer wants it to do—and more.
- Customer service records are created and checked.
- The telecommunications-based service provider creates accurate billing records and acts upon them.
- The service provider agrees to no arrangement or contractual commitment that will harm the organization.
- The customer signs an acceptance certificate and pays a bill to that effect.

If any part of the overall telecommunications-based service provision process is more crucial to success, customer satisfaction, and profit than the others, then it is this one of order implementation.

Continuing Service **11**

11.1 INTRODUCTION

A manufacturer may survive and prosper with a moderate, or even poor, level of service if the platform or device being sold is outstanding enough to ensure continuing demand. Monopolists in a service environment will also prosper within the bounds in which they are set. Consideration of the relative performance of national telcos leads to the conclusion that monopolists are less effective and passionate about the service they give than those working in competitive environments.

Telecommunications-based service providers live or die by the quality of the service that they give to their customers. Service is the very essence of the competitive propositions described in this book. This chapter is focused upon the continuing service phase of the telecommunications-based service provision value chain, which follows the receipt of an order and the subsequent acceptance of the product into service with the customer.

Continuing customer service (see Figure 11.1) is the process area that lasts longer than any of the others, so it is also the process area with the largest potential number of "moments of truth"—moments or episodes where the customer is actually using the service or is experiencing the backup that enables the use of that service.

11.2 THE SCOPE AND RATIONALE OF SERVICE

People are not always entirely rational in their approach to *service.* In the odd way that humans treat important but "difficult" issues, the subject of service is blessed with much anecdote, black humor, disinformation, despondency, and graffiti.

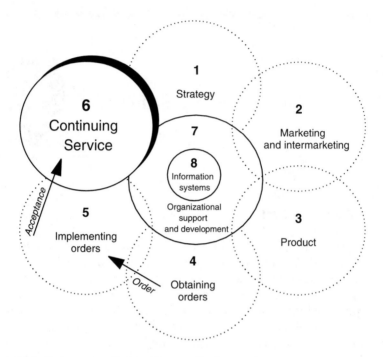

Figure 11.1   Process area 6—providing continuing service.

- "Who sold you this?" says the service engineer
- "You visited me with ten people at our first meeting; now you don't even return my calls..."
- "Ah. Then you were a sales prospect...you're a customer now."
- "We're short of resources."
- "You don't care...your attitude to service is terrible"
- "No, I can't help you...that's a network fault; call the telco."
- "I can't do that over the phone. You'll have to write in."
- "Transfer him to a customer service role, he's not good enough for sales."
- "I shall stop being your customer as soon as I am able...and I'll be telling all my friends..."

Service is important to customers on a number of levels in the mind; it is experienced in dimensions both open and more hidden within the psyche; some of the reasons for this are discussed in Section 11.3.

Customers complain a lot, but there will always be some customers who say, "Thank you. Who can I call to tell what a good job you've done for me?" This serves to illustrate the direct link, or duality, of customer satisfaction and employee satisfaction/ effectiveness.

Enlightened suppliers, be they strategists or implementors, understand the role of service and customer relationship management:

- "It costs thirteen times more to win a new customer than to keep an old one."
- "A well-handled customer complaint does more to keep a customer than a service that never goes wrong."
- "Think about customer service first...if we get the revenues, then the costs will look after themselves. No customers, no business."
- "Our customers are the best source of information we have about the product we sell...we analyze each transaction they carry out with us."

Service, whatever it represents to different people, is at the heart of the matter. What is it? What makes service good or bad? How should service be undertaken by the telecommunications-based service provider?

The effective acquisition and retention of customers for a telecommunications-based product is a holistic undertaking; it requires effective systems and procedures, well-trained and motivated people, and adequate tools and resources throughout each of the phases described through the eight chapters of Part 2.

Chapter 10 showed the crucial importance of getting the implementation right. With a good implementation, the *provide continuing service* process should be off to a good start.

Customer service for a telecommunications-based service provider is based around the call center. Before going on, it is worth referring back to Table 9.2 and its adjacent text, and to Section 10.6 and various related sections and tables.

11.3 PHILOSOPHY AND PSYCHOLOGY OF SERVICE

People know good and bad service when they encounter it, but service itself is difficult to define. The taxonomy and conceptual scope covered by the word *service* is wide and confusing.

Service is a word meaning many things to different people, and each meaning conveys different connotations. Service is an action by one party in support of another. Service is a synonym for *product.* Service may be something that happens once (also described as an *instantaneous implementation* in Chapter 11) or it may be a lasting state (*in service*) of a product whilst it is being used by customers. Inanimate objects, and even farm animals can, *be serviced* (car maintenance, hotel room cleaning, fertilization).

Service can be regarded as selfless and noble. Service brings out the best in people; many people, if not most, get a kick out of being of service to others. For many people it is a core life rationale.

But, on the other hand, perhaps in a national or cultural sense in some countries, those who provide service (servants) are commonly regarded as inferiors to those who pay them and give them orders (masters). *Service industries* can even be regarded as being inferior to more economically and socially acclaimed *production industries*.

Service means several different things at once; it is experienced and measured in many ways; it is more important at some times than at others. The service relationship can play upon some of human nature's less pleasing aspects; service activity often takes place when a customer is being frustrated in the normal use of the product and thus unable to achieve business or personal objectives. What may appear trivial to one person may be experienced as important or even vital to others.

- A television rental company told a customer that it would take two days to repair a faulty set..."But what will we do?" was the anguished response.
- "My business is entirely dependent on you restoring my mobile telephony service immediately."

Frustration is exacerbated by the feelings of helplessness and unwilling dependency engendered.

Customer behavior may also vary. Being a customer can bring out the darker side of the person, providing a stage upon which anger, pomposity, and self-righteous indignation are given legitimacy by the implied *master/servant* relationship that exists between customer and supplier. Suppliers may retaliate or protect themselves with exaggerated courtesy and *passive-aggressive* behavior.

Nobody wins arguments with customers. Yet, if suppliers give customers all that they say they want, they will also fail. There are "customers from hell" and they should be sent back there if they are not contributing positively to the objectives of the organization. In many cases, though, it was the fault of the organization for not listening and for not managing expectations in a systematic and disciplined way.

Good training, a resilient humor, a sense of being empowered will always provide a real chance for employees to achieve a turnaround and better outcome with such customers. If customers are passionate about the product, it is largely because of the benefit it gives their businesses or for the contribution it makes to their lives. They can become the most articulate of advocates.

Fundamental to good service is effective communication between telecommunications-based service provider and customer. The processes of that communication, including the brand (the position and messages about a product that people hold in their minds.) and "moments of truth" are covered in Chapter 8. Fundamental to dialogue is the establishment of clear and legitimate objectives. These are introduced in Sections 11.4 and 11.5.

11.4 BALANCING CUSTOMER AND SUPPLIER NEEDS

Any fool can satisfy customers and exceed their requirements, given enough resources and focus. Keep giving customers more and more...and charge them less and less. But is that the way to maintain a competitive business?

Most effective productive customer relationships are defined by a very small handful of factors. Obviously it is important to get these things right. But everything the organization does over and above these critical factors is a waste of resource and a source of lost focus.

> "For most organizations, the answer is to do less for customers rather than more. Just make sure that the few things you do are things that will encourage the customers to behave in the way that you want them to behave. The objective should not be customers who are satisfied and delighted. The objective should be customers who contribute to the success of the organization—by making sure the total value of what they buy exceeds the total cost of providing it. Unfortunately, a great many organizations have lost sight of the real objective. Satisfaction scores have become the objective in themselves" *(Steve Jolliffe, Chairman, B.E.M. Limited)*.

In principle, the longer the *continuing service* phase lasts, the greater the profit for the telecommunications-based service provider. But, as can be seen, rhetoric of "continuously exceeding customers expectations" is rapidly being discredited.

11.5 THE OBJECTIVES OF CUSTOMER SERVICE STRATEGY

The effective and profitable provision of a telecommunications-based product or service will depend heavily upon the successful achievement of the earlier phases in the overall process. Goals and home runs rarely result when the midfield work or the initial pitch are deficient. They never result if a playing field is marked out for the wrong game. The provision of continuing service is part of a jigsaw puzzle whose overall objective is to provide a mediation between customers and technological/information elements for the benefit of all the stakeholders.

Summarizing points in the chapter so far and other relevant factors, it can be adduced that each telecommunications-based service provider needs a customer service strategy that achieves the following seven objectives:

1. Customer is active and effective in using the product.
2. Customer is insulated from problems.
3. Customer pays, on time, more money than the service costs the telecommunications-based service provider to provide.
4. Customer receives from the product more value (to her business or life) than she pays out to the service provider.
5. Information on the customers, their usage, and the financial effects are gathered, processed, communicated, and used.
6. Proposition is improved, broadened, and deepened (value engineering, "up-sell, cross-sell").
7. The relationship is sustained over time.

11.6 PROCESS DESCRIPTION

If the objectives above are to be achieved, then the *continuing service* process must have within it effective systems and procedures, well-trained and motivated people, and adequate tools and resources.

Table 11.1 lists some of the pieces in that part of the jigsaw puzzle.

Table 11.1
Process Outline—Provide Continuing Service

Field	Attribute
Process area description	Provide continuing service
Process owner	Customer service management supported by billing, customer assistance, training, and repair

Field	Attribute
Processes and activities	Issues of customer service should be identified and addressed prior to the acceptance of an order and during the implementation phase
	Accept handover of the customer from the "implement orders" team Initiate the "customer service relationship" Provide "help desk support" Bill the customer Manage technical faults and functional problems Carry out shifts and changes Provide enhancements Measure/conduct dialogue/analyze...and act upon what is found Discontinue service
Time to complete	For practical purposes
Inputs (and sources)	A fully implemented sales order ("implement orders" process) Cash for providing the service in continuation (customers) Fault reports (from customer or from operational management who spot a problem before the customer does Change requests (customers, usually) Various elements for minor upgrades and alterations carried out within the general scheme of the original order (various sources) Requests for the contract to be terminated or amended significantly, or leads for a new sale (customer)
Outputs (and destinations)	Bills and other periodic communications related to the use or enjoyment of the product (customers) Information related to episodes that interrupt the normal use of the product (e.g., general faults, progress on rectification of customer specific faults, etc.) (customers) Information about the way customers, individually or as a whole, are using products ("create and monitor strategies," "create and manage product")

Table 11.1 (Continued)

Field	Attribute
Outputs (and destinations) (continued)	Operational inputs to "manage product," "obtain order," and "implement order" that result from each of the activities covered above
Lead process area for these information systems...	Inbound telephony: PABX lead user customer service system—inbound call handling Customer and employee training systems
Support documents	Customer service plan Customer record Records of interactions with customers—and actions/observations resulting Fault-handling documentation Bills Questionnaires Information management
Effectiveness measurement—to measure achievement of the following seven objectives: Customer is active and effective in using the product Customer is insulated from problems Customer pays, on time, more money than the service costs the telecommunications-based service provider to provide Customer receives from the product more value (to her business or life) than she pays out to the service provider Information on the customers, their usage, and the financial effects are gathered, processed, communicated, and used Proposition is improved, broadened, and deepened (value engineering, "up-sell, cross-sell") The relationship is sustained over time	Budgeted and actual cost of each product in operation Performance statistics, application response times, number of faults/failures/degrades, availability (say 99.98%), mean time between failure (MTBF), etc. Time to resolve issues, mean time to repair (MTTR) Service interface performance: help desk/service management answering times Customer satisfaction—quantitative and qualitative Profitability, usage, and other information arranged "per customer"
Risk and cost management/disaster recovery	Two types: Activities to mitigate the effects of general breakdowns, shortcomings, and disasters Actions taken to restore individual customer satisfaction levels to the point where they are likely to remain as customers

11.7 INITIATING/MAINTAINING "CUSTOMER SERVICE RELATIONSHIP"

Having accepted the responsibility for the customer from the "implement orders" team, the customer service team needs to check a number of items:

- Are the details about customer and product recorded correctly?
- Is billing set up?
- Does the customer know how to use the product?
- Does the customer know what to do/who to call if anything goes wrong?
- Would the customer like to add to or change any aspect of what he or she has contracted to buy?
- "How have we done so far?" "Anything else?"

A telephone call to the customer is probably a good way to do most of this; it is likely that an instruction book and "welcome pack" has already been sent out, perhaps with the delivery of the product.

A well-run customer service operation is likely to be equipped with a customer database and with screens and scripts that make the check call systematic, consistent, and easy to do.

Open the call with a courteous request to ensure that all details are correct so as to ensure a long and mutually beneficial relationship. Moving into the "how have we done so far?" phase of the call, avoid lengthy and numerous questions on the minutiae of the sales and implementation process. Not only can they irritate the customer and lead to low completion success, but they can also be so unwieldy that they fail to convey the vital few simple messages. A simple questionnaire with four open questions will elicit all that is required to the level of detail self-selected by the recipient. If there is a material problem with, say, the service provider's courier delivery service or with the confusing wording of the instruction manual, then it will certainly be mentioned by a significant number of customers.

Example from Section 7.6.1

1. *What do we do well?*
2. *What would you like to see us do better?*
3. *What has changed over the last six months?*
4. *Will you or anyone else you know want to buy more of our product in the near future?*

These setting-up and checking activities are largely neutral in terms of the relative psychological positioning of customer and customer service agent. The agent needs to be pleasant, confident, and well-informed. The agent must be a good listener—providing "active listening" prompts as required—and must be sure to record the results sensitively and accurately. Active listening prompts might be used to probe discreetly for particular possible problems if they are suspected.

Customer service agents will appreciate being given the time, authority, and tools to set things up well, especially so if they know that this customer is to be their continuing responsibility. If the agent is given this early opportunity to get to know a customer in a stress-free interchange, then the ability for agent and customer to interact when things *are* going wrong is strongly enhanced.

For a large customer, a "customer service plan" will be a formal and detailed document set to cover all of these issues in a convenient format (see Section 11.13 and Section 7.6.2).

11.8 PROVIDING "HELP DESK SUPPORT"

Help desks come in several forms. *Help desk* may be a label for any of the following:

- Fault reception;
- Billing and payment inquiries;
- Contract inquiries;
- General product information;
- Specialist product information;
- Application/user help;
- Software snags and bugs;
- Hardware remote diagnostics and restoral.

A challenging starting point in defining a help desk is that "its staff will find and pass back information in respect to any question that they are given." In practice, a major telco's product and services help desk is unlikely to receive inquiries about motor tires but, if it were to do so, then the staff would still be able (if they chose to do so) to direct the caller to somewhere else that could provide the information required. In practice, callers will select the most likely source of help from a directory or from product literature, or perhaps by calling a general number and being routed via an interactive system (such as IVR). In any case, there will be a high level of subject matching once the caller has reached the help desk agent.

Help desks learn and begin to network with each other.

Example: BT Product and Services Help Desk 1988, London

"When the two of us were assigned the task of taking product inquiries, we didn't really know what to expect. When the first calls came in we didn't know the answers to the questions, so we had to research them and call back with the answers. Probably more by accident than design, we kept a detailed record of each inquiry, where we found the answer, and what we did about it. Soon, we found ourselves with shelves and files of information contained in manuals, odd faxes, memos, product specifications, brochures, and price lists. We began to build up a repertoire of answers to frequently asked questions. We became better and better at answering questions; more and more people began to use us. Our clientele came from inside and outside the corporation. Telecommunications consultants started to use us as an information source. Four months on, we had taken on three more people. We had a breakthrough when we took on a Ph.D. student on vacation from her program of study in librarianship. She showed us a lot about cataloguing information so we could store and retrieve it efficiently. It so happened that a BT pensioner came and worked with us for three months; he was able to talk us through issues on older products and provided us with knowledge and insights from the past that had slipped away from our generation. Then something else started happening: product managers began to come to us for information, too, rather than us just going to them. We told them about the customers asking about potential features; we told them about the difficulties some people had in using their products. Interestingly, our pensioner spent more time calling around his contacts than he did searching through the books. By this and other means, we found more and more help desks out there in the world and more and more of them found us. Our Ph.D. student also kept logging on to something called the Joint Academic Network/Arpanet to bring back knowledge from all over the world...maybe one day that network could really grow into something big and widely pervasive..."

And so it did. Most of the information in the world about telecommunications products can be brought to a single terminal, if you know how to do it. Help desks know.

The people who work on help desks tend to be intelligent, curious, and helpful. They think laterally and are disciplined in execution. They acquire knowledge and infer widely from it. To start a help desk, find a person with those characteristics, give them a telephone and a desk, and then let it run.

11.9 RELATING HELP DESKS TO CALL CENTERS

Section 9.5 provides a description of a call center, characterizing it as " at the heart of the processes and resources deployed to handle the volume customer and, as shown above, to handle the high volume or simple transactions for other sections of the market (such as major customers)."

Help desks are not automatically synonymous with call centers, but the functionality of each makes the other operate. The larger and busier a help desk becomes, the more likely it is that call center functionality and processes will be required to make it work effectively. A call center may only operate to provide a single simple function (e.g., recording names and addresses on a database and sending requested product brochures), but as soon as the range and complexity of the inquiries increases, then the greater "intelligence" of a help desk is required.

At the other end of the spectrum from the knowledge-rich help desk, the agents unfortunate enough to handle single-transaction calls will be hard to motivate, will soon communicate their boredom to callers, and will move on. They are likely to leave a trail of havoc and half-correct, misspelled customer records. They are working in the information age's equivalent of a sweatshop or coal mine.

Wherever possible, machines should be given this type of work to do. Interactive voice response operates on touch tones or using voice recognition. The latter comprises small-, medium-, and large-vocabulary applications. The world's stock of phonemes (parts of words and phrases) analyzed and captured on voice recognition systems is increasing rapidly. They tend to roll out as American English, British English (of which there are 19 regional variants), Spanish, Mandarin, German, European French, and so forth. Reliability for large-vocabulary applications is already over 90%. As already described, IVR systems and their WWW equivalents enable callers/computer keyboarders to interact directly with host computers providing input, inquiry, and output mediated at will between tones, text, and speech.

11.10 BILLING THE CUSTOMER

11.10.1 Introduction

For convenience, both design and operating issues are handled in this section.

Whole books are written on billing. Telcos have many legacy systems. New startups build their own on a PC, and they work but won't scale. All around the world the search goes on for a perfect system. No two telecommunications-based service providers billing requirements and solutions appear to be the same.

As a rule, billing systems are expensive to maintain, take too long to put in place or enhance, are tricky to integrate, and fail to live up to the claims of their suppliers.

There can be little doubt that the billing system will be a major determinant of the product's economic success and its acceptability and performance in the domain of customer service. In practically all circumstances, it must be developed and tested before a product is put into service. A billing system that is anything less than excellent will constrain the freedom of product managers to realize the full potential of a particular product or market.

Billing systems must operate effectively within the overall information architectures exemplified in Chapter 14. Billing systems are the rightful half-brothers of customer support systems. Regrettably, they both can end up as orphans. Customer support systems are often not designed in line with or in step with the central product concept. Billing and customer support systems are not seen as challenging and exciting projects—but IS specialists and managers who work on billing systems end up as scapegoats as often as not.

11.10.2 Essential Components of a Billing System

The essential components of a billing system are as follows:

1. Collectors of call or transaction data—call detail records (CDRs), transaction tapes from suppliers of information elements, journal entries from transactions like hardware purchases, and so forth;
2. Tariff table—what to charge in particular circumstances for particular transactions;
3. Rating engine—applying 2 to 1 and storing the result by customer record;
4. Bill format table;
5. Periodic conversion of results from 3 into a bill in the format drawn from 4;
6. Printing of hard copy bills and stuffing into envelopes/or EDI equivalent;
7. Procedures, tools, and people to receive payments and apply them to corporate bank account;

8. Help desk to answer queries;
9. Credit control and enforcement of payments;
10. Linkage to corporation's customer service systems and main accounting systems.

11.10.3 Will it Work?

As for any IS development, the path to success is to follow an effective systems development life cycle using a robust and proven statement of requirement and a robust and proven billing system supplier. Some killer questions to ask are as follows:

- "Has the potential billing operations team reviewed the user spec?"
- "How, precisely, does the billing system design map on to our corporation's data model and on to our open systems architecture?"
- "In how many places has the system you seek to sell us already been implemented...may we go and see some reference sites, please?"
- "Where do you already have this system working with the same combination of switch, computer hardware, signaling system, system software, database management system, operating system, and protocols?"
- "Is the source code written in a (national) language you understand?"
- "Would it be possible to use a bureau service?"

11.10.4 Enhancements and Priorities

It is in the nature of billing systems that they require constant enhancement. In Section 13.7, a simple but effective approach is given that will deliver against agreed objectives and will foster concord amongst the system management team, the IS steering group, and users.

11.11 MANAGING TECHNICAL FAULTS AND FUNCTIONAL PROBLEMS

Customers will undoubtedly be provided with a warranty, and thereafter they may be able to have fault rectification service, equipment loss protection/provision of replacement equipment, and so forth, either under an insurance policy or on a per-occasion charging basis. The former arrangement represents an excellent value added cross-sell opportunity and it provides a mechanism for remaining in long-term contact.

Faults and functional problems can be described at three levels:

- Level 1: Online, immediate fix from help desk;

- Level 2: Specialist repair/workaround or unit replacement;
- Level 3: Workshop repair/longer timescale software bug fix.

The procedures or components of a typical fault rectification system are as follows:

- Management:

  1. Receive call—bring up customer record and fault logging screen.
  2. Log call and record appropriate information about the call, the presumed fault, and the customer.
  3. Evaluate the problem in consultation with the customer. Carry out remote online hardware/software diagnostics.

- Level 1:

  1. Provide telephone advice/carry out remote on line rectification.

- If fixed, go to 11. If not fixed, go to 5.
- Level 2:

  1. Assign the fault to a specialist within the telecommunications-based service provider organization, to an outsourced repairer, or to the subcontract supplier of the relevant information or technological element (Level B).
  2. Carry out repair.

- If fixed, go to 11. If not, go to 7.
- Level 3:

  1. Pass to workshop, software support, and so forth.
  2. Continue till fixed or abandoned.

- Management:

  1. Advise customer regularly of status and progress to resolution.
  2. Escalate problems to senior management according to preset elapsed time or customer/product significance parameters.
  3. Advise customer that fault is cleared and get agreement to that fact.
  4. Record all fault data and pass to product management for assessment and overall rectification/enhancement action.

Effective tools and systems need to be provided for the management of fault rectification; these will include call center functionality, terminals, modems, customer records, repairer/workshop tools and spares, testers/probes, mobile telephony, and so forth.

Information will be needed from manuals, online help screens, product specifications, and other similar sources. People fixing the problem need access to help desks and specialists. Contact/call out information is needed.

Telecommunications-based service providers should maintain information on interactions with customers as a chronological log. How much more acceptable it is, in dialogue with a customer service agent, to have information presented back to you over the phone in the same manner in which it resides in your mind.

Virtually all of the 12 functions outlined above can be outsourced in some way or another. Modern telephony and agile distributed databases enable call reception and fault management for local centers to be dropped back out-of-hours to central points. With falling telephony costs and global markets, the central point is quite likely to be in another country. With a fully global corporation, it is increasingly common to have three or more fault management call centers situated in different time zones. Specialist services are available to receive calls in one country and answer in the language that accords with the origin of the call. Holland, for example, is a country that specializes in multilinguistic capability. Some telcos offer multilingual operators who will stay on the line and provide translation support.

The ideal service/fault receptionist is calm, a lateral thinker, responsible, knowledgeable, and empathetic with the customer—seeing the problem from the customer's viewpoint as well as that of the telecommunications-based service provider.

Engineers who go to site will benefit from similar skills, but must above all be knowledgeable and practical. The appearance, behavior, and general demeanor of service engineers is a material factor in the projection of the brand of the service provider.

An acid test check of service effectiveness is to cross-check the records of the customer and of the customer service management function: "What does *the customer* perceive to be outstanding at this time?"

11.12 DOING SHIFTS, CHANGES, AND ENHANCEMENTS

A service center should be equipped and its people trained to carry out whatever is required, including the sale of minor enhancements. It may be appropriate to pay commissions or bonuses for these extra sales. Views vary greatly on this last point.

11.13 LARGE CUSTOMERS, IMPORTANT CUSTOMERS

It only remains to say that all customers are not equal. Some customers will be prepared to pay more for a premium service, or for an insurance policy. Some customers will need to have out-of-hours service or rapid response. Some customers, especially heavy users ("frequent flyers"), will expect the telecommunications-based service provider to repay loyalty with privileges and special offers.

The telecommunications-based service provider should consider the efficacy of managing schemes like these. Affinity marketing is a sine qua non if a long-term relationship is sought. The best way to find out about them is to look at the way airlines do it.

Large customers will expect specific arrangements, particularly the assignment of named customer service agents, help desk contacts, and customer service management. Matters such as these should be agreed and documented in a *customer service plan.* It is important to meet with such customers on a regular basis. It is vital that customer service and other issues be recorded and addressed. Do not forget to tell the customer when matters have been fixed.

All customers may not be equal, but all are important. In Section 11.7, the process of initiating the customer service relationship with a welcome call was described. If the customer/agent relationship is formed in that way, then the agent will remember the customer for next time. A team of three or four agents is probably able to provide a high level of warm personal service over the telephone to 1,000 customers or more. If it costs in, then do it.

11.14 HANDLING INBOUND CALLS AND OTHER MECHANISMS

11.14.1 Roles and Responsibilities

Clearly, the service effectiveness of a telecommunications-based service provider will be heavily dependent upon the effectiveness of inbound call handling.

Defining, providing, and monitoring the socio-technical systems for handling all inbound calls, messages, letters, and personal visits by customers is the responsibility of the VP for customer service so that a uniformly high level of service is provided. Inbound cash handling and billing processes should also be included.

The VP for customer service is not alone in this task. The call center provides an integral part of the systems for both marketing and for obtaining orders. Common "look and feel" is an important part of branding—the purlieu of marketing/intermarketing—but the service level standards and the process integrity are the responsibility of customer service. For convenience, the architecture and use of the call center was described in Chapter 9, but many of the underlying systems are the responsibility of customer service. Cash handling is a key responsibility of accounting (within *organization support*), but it is customer service that must ensure the operation is effective from the perspective of the customer and the customer service systems.

More details of roles and responsibilities for telecommunications-based service provider systems are given in Chapter 13.

11.14.2 Telephone Answering and Message Management

Not everyone is in the call center environment. There are three phases of handling incoming communications as an organization grows:

1. Central administrator/receptionist for a single team;
2. Direct to the individual's desk/phone, backed up by a team administrator;
3. Full call center.

When an organization first starts out in life, all the people sit together. Calls and other inbound communications are handled effectively by these people and/or a team secretary on their behalf. Everyone knows the role, current issues, and current whereabouts of everyone else.

Later (at about 10 people), the number of internal interrelationships/communication paths reaches an astonishing five million (11 factorial). The team and single receptionist system will break down somewhere between seven and 20 people, and should be replaced by phase 2 arrangements.

The simple solution would seem to be a break up into subteams. However, this alone will not work. At the same time that teams have formed and administrators are put in place, the individuals within the team should have developed self-reliance for telephone answering. Assuming the switch and other equipment is suitable, the normal individual approach works as follows:

The caller is encouraged to use direct inward dialing/direct dialing in (DID/DDI):

1. *Employee* answers;
2. If not, phone *diverts* to either 3, 4, or 5;
3. *Alternative number,* at which the employee is working (including mobile phone or home);
4. *Administrator,* or colleague;
5. *Voice mail* (which *must* have a message offering 4);
6. If none of that works, the call goes to *reception/operator.*

The responsibility for and control of answering remains with the employee. Voice messages can be retrieved at any time and from any place. The relevant administrator is on hand if rediversion is required. This is a very simple implementation of personal telephony and an elegant way to implement one aspect of location-independent working (also teleworking). The method described above is clearly adaptable to meet "boss-secretary" needs.

These arrangements work extremely well for an environment where each call is different from the last, where they are characterized more by complexity than by volume, and where a response by a particular person is needed within, say, six working hours. For time-critical communication, some form of live answering is essential.

Where teams are very large, where they cover a wide time slot, or where the incoming calls are tending to be intensive and repeated transactions, then an increasing level of call center functionality needs to be introduced.

11.15 DISCONTINUING SERVICE

All customer contracts come to end some time. It is important to predict when they will do so, for it will both aid financial and resource planning and (more creatively) will give early warning to the telecommunications-based service provider and an opportunity to save some of the terminations. A saved disconnection of an existing customer is usually more profitable than a new connection. Cost-benefit analysis indicates that the telecommunications-based service provider can invest relatively heavily in a dedicated team to prevent disconnections occurring. The reasons for disconnection (or potential disconnection) must be investigated rigorously and eliminated or ameliorated as much as possible. The *saved disconnection team* will become knowledgeable and adept, and these qualities can be deployed "backwards" into the customer base to predict potential problem customers before they take action for themselves. Consider the following example.

A saved disconnection team identified that high-usage mobile telephone customers ($1,000 per month) changed their equipment around 22 months. They would be stimulated into this by the advertisements they saw in newspapers or heard on local radio. They would follow up on these advertisements and not inform the existing provider until the new purchase had been made. Using this information, they scanned the customer database for such customers and arranged to offer them a new handset after 18 months. The problem was significantly reduced.

They then went further and offered a deal to high users whereby the existing phone could be reconnected free for a family member, friend, or business colleague and allowed $100 of free calls. The reasoning was that the high user would make enough calls to the new user to offset the net cost to the service provider of this "gift" to the latter. The latter himself, meanwhile, was stimulated into the phoning habit for himself. Everybody gained something.

Disconnections should be handled with the same care and courtesy as any other transaction. This customer may come back; there may be an opportunity to sell another product.

11.16 CONCLUSIONS

Good customer service requires effective systems and procedures, well trained and motivated people, adequate tools and resources. This chapter has identified how good service can be provided and set that in the context of the psychological and practical issues that arise.

The processes of service are both *rectangular* and *fuzzy* (see Section 5.4.3), the semantic basis of service is both rational and emotional. Since it could cost 13 times more to win a new customer than to keep an old one, it is worth investing heavily in the continuing service phase. Customers can get bored; they hate to be ignored; they don't even mind if things go wrong, provided they are put right quickly, pleasantly, and effectively.

A well-handled customer complaint does more to keep a customer than a service that never goes wrong.

The continuing service phase is the part of the customer relationship that lasts longest and the one that gives the greatest opportunities to observe customer behaviors and assess which are most profitable and to be encouraged. Powerful new computing techniques give the capability to take all of the resulting information and enable a virtuous spiral of improved performance, growing customer satisfaction, enhanced relationship, and broader and deeper use of the telecommunications-based service provider's products. Profits rise.

More service than is required is not better than good service. By all means exceed the expectations of the customer, but only to achieve the virtuous spiral of profit over time. Customer satisfaction ratings are not an end to themselves. Employee satisfaction and customer satisfaction go hand in hand. Notwithstanding the discipline and focus of economic calculation, there are clear benefits for these two sets of stakeholders as the virtuous spiral is achieved. Given that tendency and a reasonable degree of common sense, it is difficult to see how stockholders and all the other stakeholders will not benefit significantly, too.

Organization Support and Development 12

12.1 OVERVIEW

An organization that is not supported, well-led, and developed in line with its business growth or changes will surely erode its customer, employee, and stockholder satisfaction, eventually to the point where they will begin to depart and the business will start to die.

The majority of the issues are common to any business and most of this book's readership will be aware of many of the issues already. Despite being aware of the issues, however, many entrepreneurs do ignore the essentials, especially as the organization grows.

In discussing the balance between these considerations, this chapter concentrates on just two aspects:

1. Listing the generic processes of organization support that are essential to the functioning of a telecommunications-based service provider;
2. Identifying issues and providing a few practical tools for organization development and leadership that need to be considered in telecommunications-based service provision.

An abbreviated process area outline table is given for each of the three, followed where necessary by brief text and some diagrams.

Organization support impinges upon all the front-line processes, as shown in Figure 12.1.

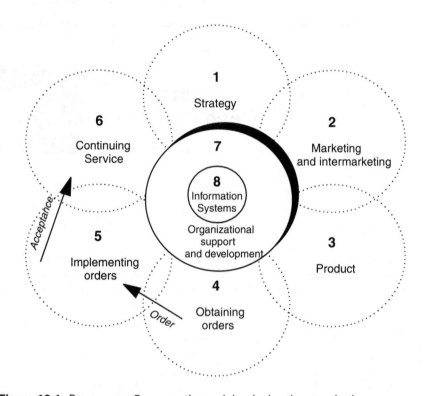

Figure 12.1 Process area 7—supporting and developing the organization.

12.2 GENERIC PROCESSES OF ORGANIZATION SUPPORT

Some of the activities shown in Table 12.1 may be handled in one of the other process areas on behalf of all (e.g., logistics may be handled by *order implementation*).

12.3 ORGANIZATION DEVELOPMENT

The processes of organization development for a telecommunications-based service provision are no different and no less vital than those of any other complex, fast-moving business. Figure 12.2 gives a framework tool that should be used in association with the strategy tools described in Chapter 6.

Table 12.1
Process Area Outline—Organization Support and Development

Field	*Attribute*
Process area description	"Organization support"
Processes and activities	Financial accounting, cash management, taxation Company secretarial Commercial
Lead user for these information systems	Financial accounting systems including automated payment systems, banking, etc. Individual customer contracts Personnel systems, payroll Quality management system, procedures Building, vehicle records; fixed asset register Audit and security, disaster recovery, insurance

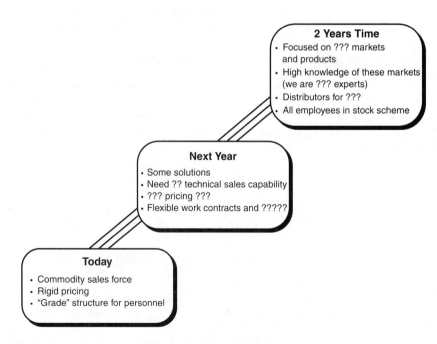

2 Years Time
- Focused on ??? markets and products
- High knowledge of these markets (we are ??? experts)
- Distributors for ???
- All employees in stock scheme

Next Year
- Some solutions
- Need ?? technical sales capability
- ??? pricing ???
- Flexible work contracts and ?????

Today
- Commodity sales force
- Rigid pricing
- "Grade" structure for personnel

Figure 12.2 Organizational development framework.

12.4 PROCESS MANAGEMENT—*QUALITY*

12.4.1 Conflicting Drivers

A telecommunications-based service provision is likely to be a fast-moving and often changing organization. It will also probably be required to register under a quality scheme such as the international quality standard ISO 9000. Anyone who has been through this will know both the pain and the time that this takes. By the time processes are documented in detail, the world has moved on...even if there were time to document them in the first place. How can the circle be squared?

First off, it seems inevitable that everything will be done by a few people who all talk with each other all of the time. Processes aren't written down. By the time there are 20 people, there are millions of possible communication paths amongst them. Those who have been there a while know how to make things work, but those who have not are left helpless. If you really want to know how things work, then "go and ask Fred" (see Section 5.4.2).

So are process documents actually worth having? Are all our main processes the same in character? Do they all need documenting the same? Will people read all this documentation, even if it does get written?

12.4.2 A Simple Quality Management System

The document architecture shown in Table 12.2 will provide a simple structure for most of what is required. The architecture is totally consistent with ISO 9000 requirements. It also includes job descriptions for each job and person and a "person specification" associated with each job to use for recruitment and development purposes.

Table 12.2
Simple Quality Management System/Documentation Structure

What It Is	*What It Contains*	*Where You Will Find It*
Process and quality manual	Introductory section Organization structure/role definitions Description of each of eight process areas Title/objectives/goals Flow chart Ownership	Master document held as read-only file. Small-size landscape hard copies distributed periodically to all company members

What It Is	*What It Contains*	*Where You Will Find It*
Process and quality manual (continued)	For each process (e.g., "bid management"): Purpose, owner, inputs, outputs, activities, resources, reference to procedure documents	
Process area index	List of procedures, work instructions, and forms (i.e., numbers, titles, and issue status)	Master document for each procedure held as a read-only file on the network. Controlled and uncontrolled hard copies produced only as necessary
Individual procedures grouped into process areas	Formal description of the activities that form building blocks for the processes above (e.g., "customer proposal document") Title/number, process area, scope, forms and references Responsibilities—details	As above
Individual work instructions	Detailed tasks, instructions, checklists, instructions for use of particular systems or tools, etc. In similar format to three procedures (e.g., "telemarketing contact checklist or scripts")	As above
Forms	Documents in a mandatory layout, which is used to implement and also record the procedures and work instructions ("fax header sheet,," "engineering change request,," "training evaluation form")	As above
Guidance documents	Ideas and guidance to aid development and understanding of the formal documents (e.g., head office procedure document not exactly right for the subsidiary, white papers	Text books

Table 12.2 (Continued)

What It Is	What It Contains	Where You Will Find It
Quality plans	Documents how quality will be achieved for a particular product, project, or service (e.g., may be in a bid document)	As above, associated with documentation of subject
Job descriptions and person specifications	JD: JD number/job title—JD subnumbers...related to... name of each individual jobholder, purpose of job, reporting relationships, responsibilities showing % of whole and written to include functional and personal development objectives for the current year. PS: PS number related to JD number above, job title, experience required, formal qualifications, skills, personal characteristics, travel, and other requirements	Held on central server and as hard copy file. Easy access (read only) and hard copies for each individual and for others who require them

12.4.3 Organization Charts

Traditional organization charts are strong at showing hierarchical relationships. Experience shows that the use of twin diagrams as in Figure 12.3 provides a clearer picture, especially if all of the names can be shown on one sheet. The distinguishing feature of this twin scheme is that it is based initially on process areas rather than hierarchical teams. People are likely to appear several times over.

12.5 LEADERSHIP

Leadership is conducted at many levels—intellectual, emotional, and even spiritual. Leadership is a mixture of qualities, skills, and tasks. This section is concerned with just tasks.

The role of the telecommunications-based service provider's leaders is to act as compass, energy source, and architect.

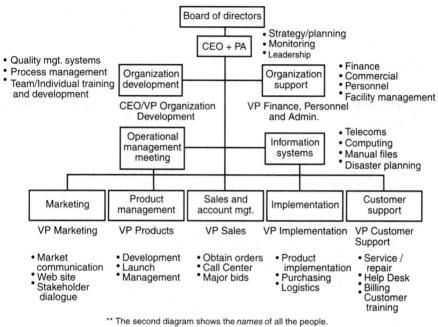

Figure 12.3 Two-part organization chart.

As a *compass*, the leaders must look ahead and make sense of the environment so as to spot the opportunities and dangers and then ensure that all of the constituents are directed in the same way. Remember how at school a piece of paper was piled with iron filings and then shaken gently but continuously above a magnet? Gradually, all of the iron filings are pointing in the same way, heading for the pole. Those who work in a telecommunications-based service provider are frequently faced with new situations and need to make decisions on the spot—especially if they are customer-facing. Such people need a clear understanding of direction and destination. Tools such as the strategic staircase work well for people in this regard (see Figure 6.3).

The *energy source* task is conducted by listening to concerns, providing encouragement, and making resources available at the right time. There are whole libraries devoted to motivation, resource planning, and all the rest.

To lead the process of stakeholder dialogue is to lead the organization. In Section 7.6, one task of the CEO was stated to be "...to give each group what it wants. In practice, this implies managing the expectations of each in line with reality. If there is dialogue between groups of stakeholders, each with its own knowledge resources and paradigm, then priorities will be easier to assess and the *reality* may move for each party. Effective management of this dialogue will make the difference between a perception of success and a perception of failure."

As an architect, the leader must ensure that each person and team understands what is expected of it—and what may be expected of others.

Just five tools will provide much of what people seek from their leader:

- A vision of the future coupled to a realistic view of the present (Where are we going? Where are we now?);
- Up-to-date organization charts (Who does what?);
- Coherent job descriptions with tactical objectives (What is expected of me and of my team?);
- Regular reports of status and action plans—both face to face and documentary (What precisely is happening? What precisely is meant to happen next?);
- Time to talk...because nothing is simple.

12.6 CONCLUSIONS

The objective of this chapter has been to make the telecommunications-based service provider more aware of issues of support that are often put to the back or forgotten. Initially, support *can* be ignored. At startup, it is survival—not administration—that is the dominant driver. But to continue in that state beyond the minimum necessary period is to see the organization slow down and eventually stop.

The greatest of products and the most inspirational of leaders will be negated in time if no support structure, processes, and resources are put into place and accorded a suitable level of priority and recognition. There are special issues that confront high-tech businesses such as telecommunications-based service providers. They are in a highly competitive business; they are probably startups in one way or another; they will grow quickly and pass rapidly through stages of development whose nature is to cause stress for the people concerned. Out at the leading edge in their competitive marketplace, they will have individuals who have to, and do, remain highly focused on their primary task. These valuable people won't do their expenses on time and may eventually end up overrunning their overdraft limits; they may forget to have their cars serviced; they will lose things. These people will irritate the support team. The support team will probably be underresourced and undervalued. Beware of the eventual consequences.

Managing Information Systems

13

13.1 INFORMATION SYSTEMS: PROCESS AREA DESCRIPTION

The approaches, applications, and systems to achieve the ends required in each process area are described over the seven preceding chapters. systems. In this chapter, the eighth process area is described—the management of a telecommunications-based service provider's information systems as a whole (see Figure 13.1).

Imagine trying to achieve outstanding levels of success as a telecommunications-based service provider without the right information and without the right operational, management, and support resources.

Can you imagine not being able to move information easily from one place or person to another? Can you imagine not having customer information available when that customer calls, or visits the Web site? Can you imagine an information-based product on one hand with no linkage to the business systems on the other—no billing, for a start? Can you imagine denying the organization the power of Internet-based market research and customer database management systems? Or a telephone system that is always busy, goes unanswered, or directs calls to the wrong people? Imagine closing down outside the hours of 9 to 5?

There are plenty of businesses that continue to get away with performance like that, but they will not be among the success stories or even the survivors of the telecommunications-based age.

Each person or team is *enabled* by information. "Decision making has to be taken at all levels and in all areas to respond to the challenges and opportunities that an organization encounters." (Ken Fairclough and Rob Taylor, 1995, Unisys Corporation); decision making requires the acquisition, processing, application, and publication of information of all kinds. The telecommunications-based service provider operates in a fast-changing environment; if the mean time between decisions is greater than

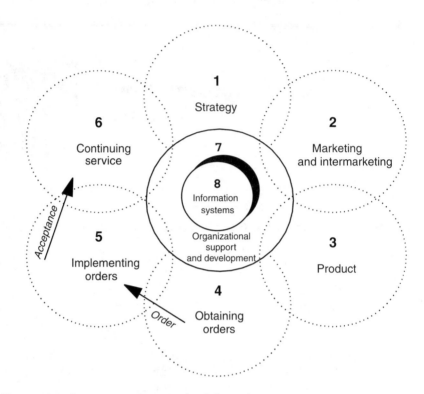

Figure 13.1 Process area 8—managing information systems.

the mean time between changes, then the enterprise will be unresponsive and eventually obsolete.

But the opportunities for excellence are there. Communication between customer and service provider has been enhanced by improved telecommunications, call center technologies, and so forth. Customers can self-serve their information needs using interactive transaction-processing systems such as IVR and the WWW. Improved computing and information management techniques, such as hub-and-spoke architectures, and improved database management systems mean that information is flowing up and down and around organizations quicker and more directly than ever before. Guidance and rules can be held online, especially by use of case-based reasoning systems, sales scripts, predesigned electronic forms, and configuration tools.

Telecommunications-based service providers can operate with an enhanced exchange of information and accessibility to information. People in direct contact with customers are better equipped to supply information and they're being empowered to make decisions on the spot. Our organizations are getting flatter.

What it takes to achieve effective Information systems management is handled in four main parts: objectives, architecture, applications/systems, and development/operation, as indicated in Table 13.1.

Table 13.1
Process Area Outline—Managing Information Systems

Field	Attribute
Process area description	Manage information systems (IS)
Process area owner	IS
Other actors	People and teams in all process areas use the information systems developed and managed under this one
Processes and activities	Set business objectives for IS Define scope of IS and architecture to match Assign applications/systems to relevant process areas Develop and manage systems
Time to complete	One month for strategy, 1 to 9 months to develop and introduce systems
Inputs (and sources)	Business requirements User requirements Funds Technical knowledge Supplier information
Outputs (and destinations)	System specifications Standards Project plans User training/instructions
Support documents	IS strategy Project plans
Information systems lead user for...	Information systems administration, assignment of equipment, licenses etc. PABX call plan etc. IS supplier information
Effectiveness measurement (measured by those receiving the outputs)	System performance against formal specification and vis a vis user expectations Uptime/downtime Cost Timeliness

Table 13.1 (Continued)

Field	Attribute
Risk and cost management/ disaster recovery	Formal procedures required, rigorously enforced Security and back up of data Firewall on Web site Password control as required Offsite backup Data and system archiving offsite Escrow copies of supplied software Telecommunications backup—multiple routes into site—from different telco switches (central offices)

13.2 WHAT HAS TO BE ACHIEVED?

Six critical success factors for the design of a telecommunications-based service provider were identified in Chapter 5. In Table 13.2, some IS implications are considered as an illustration of considerations to be considered in scope and architecture.

Table 13.2
TBS-P Critical Success Factors and IT Implications

Critical Success Factors for a Telecommunications-Based Service Provider	Example of Enablers and Contributors Powered by IS
Strong sense of direction to guide it through change	Continuous multilevel contact amongst stakeholders (voice mail, e-mail
Sound competitive proposition that is well understood	Easily accessible and attractive marketing and intermarketing material Easy to use information products Customer self-service (IVR, WWW) Built-in customer documentation Pricing and configuration built around case-based logic
Speed in taking products to market	E-mail Conferencing between locations Messaging between time slots Multiple time zones and location-independent working Iterative development of product on the network

Critical Success Factors for a Telecommunications-Based Service Provider	Example of Enablers and Contributors Powered by IS
Skilled, committed people empowered to work for their customers	Training Scripts Accessible and up-to-date product and pricing information Fewer, but more accessible, managers to handle escalations PC/terminal on every desk
Systematic but flexible processes and tools that save time	Scripts and guidance for telemarketing/sales Common application software Properly featured fax machines, high-speed photocopiers, etc.
Superior teamwork clear roles and responsibilities	Group working systems—E-mail Work management systems Organization/quality manual on the network Job descriptions on the network

13.3 INFORMATION SYSTEMS COVER BOTH THE PRODUCT AND THE BUSINESS

There is a distinction between two types of product and service:

- *Type 1*: Products are manufactured, delivered to a customer's site, and then, if necessary, repaired or serviced there. In Type 1, information comprising the product itself and information to run the business are quite *separate* (although there may be some overlaps—for example, where remote diagnostics and configuration can be carried out).
- *Type 2:* Products or services that are provided from the supplier as a continuous process—where interaction with the supplier or supplier-located systems is an inherent part of the delivery. In Type 2, the information systems are as one. The product or service, its provision, and the business processes concerned with its support are tied together within the product architecture. Even in Type 2, it will be helpful to keep *third-party information that is sold as a product* "clean," manageable, and unaffected by other information structures and systems. It could be disastrous, say, for a virus brought in by a careless employee (or a disaffected one) to infect the sold product via the administrative word processing system. But the "product," as a whole, is not just the bought-in information, but is the

packaging that goes around it of sale, delivery, help desk, and billing. In many cases, the bought-in item is a component to be combined with other similar elements, or only part of it may be used in the final telecommunications-based service provider product. In many cases, the product bought in has to have value added in the form of manipulation, transformation, or storage. Finally, the information bought in (or otherwise obtained) may have to be held on a database and accessed repeatedly as part of a service provided through a help desk agent or via interactive transaction-processing devices connected to the Internet or the public switched telephone network.

Thus, it is important that the information architecture of a telecommunications-based service provider recognizes both the required links and separations between *product* and *business system.*

13.4 HUB-AND-SPOKE—NEW AND LEGACY SYSTEMS

Something has changed in information management. It is the advent of hub and spoke architectures. Consider the following:

Johan Vinkier in *The McKinsey Quarterly* (1994, Number 4) painted the picture of an airline that kept adding more and more routes, with different aircraft, systems, crews, and schedules, all linked in to all of the other systems, routes, and schedules till the whole system became uncontrollable. Costs and customer satisfaction hit the rocks. Then the airlines discovered that they could regain control by routing everything through a single hub. The hub provided a connection from one service to another—passenger and baggage transfer systems, timetables, and conditions could be integrated. Airlines could now offer as many "city pairs" as they had aircraft to fulfill them. All they had to do was connect another destination to the hub. Passengers flowed easily through the hub.

Major organizations that operate in a transaction-intensive environment generally find that they have allowed the development of legacy systems that do not talk to each other, that have been delivered to different technologies, that have the same information over and over (but inconsistent), and that cannot be altered. In addition, links that have been developed over time to overcome these problems have become unthink-ably complex. The system finally grinds to a halt under its own weight of overhead.

Classic approaches to information system development have left IS managers with a range of unpalatable approaches: from keeping with what you have, upgrading it, and interconnecting with existing systems to starting again.

Consider for example that some years ago, it was common to see the desks in a telco fault management center equipped with multiple termi-nals, each operating on the end of a different system. Nowadays, there will be one terminal that is linked to all of the systems. Telcos used to take weeks and months to provision a new connection or a private circuit to customers. Now, they can do it in days or even hours and minutes. Often now, it is even the customer who self-provisions bandwidth on demand.

See, also the example for a European telco in Section 10.4.

The start of *call center* approaches acted as a powerful reason for change in information management. Now it was necessary for all of the information to be brought to the desk, manipulated, and used in the course of a single telephone call. As soon as one organization speeded up, the others had to follow suit. What many have done is to follow the airline analogy and adopt a hub-and-spoke approach to information management.

There are few large established organizations that do not have an extensive range of legacy systems. Virtually every telco in the late 1990s is still struggling with multiple billing systems. One global group is said to have 126. Provisioning, billing, repair, and customer support are all hopelessly divorced from each other. Many applications are fed from the basic data elements. Applying the hub-and-spoke approach, there are sys-tematic methods of identifying cross-connections and, moreover, of bring-ing a new richness to existing of these applications through an *online transaction platform* (OLTP) to a common user interface. Help from a spe-cialized information systems provider is required when constructing or reconstructing huge information systems. The linkages connect "request-ers of information" to "enterprise applications and data" via "connectiv-ity middleware" (hub-and-spoke). The bleak alternative to plotting the interdependencies properly is to switch systems off and see who squeals.

The impact of difficulties with information systems was reflected in the first few words of this chapter. But rigidity and nonperformance of telco systems represent an opportunity space for the agile telecommuni-cations-based service provider. Is there a way of going into service with-out indulging in a complex, expensive, long-winded, big bang approach to IS?

13.5 APPLYING THE LESSONS TO T-BSP IS ARCHITECTURES

The processes of telecommunications-based service providers, described in detail in preceding chapters, are the building blocks of delivering the competitive proposition. It is to these processes that the IS designer must turn to achieve success. In Section 8.7, the functions of a telecommunications-based service provider were introduced as they relate to the customer interface, as shown here in Table 13.3.

The interrelationship of processes and systems can also be shown by taking the established process area diagram (the value chain of the telecommunications-based service provider) and extending it downwards into a third dimension. The purpose of this diagram is to show the picture as a whole; referencing back to the process outline table in each chapter will show the relevant associations more directly.

A telecommunications-based service provider is there to provide a telecommunications-based link between selected customers with the providers of information and technical elements.

Figure 13.2 shows that customers interface with the telecommunications-based service provider's processes via wide area networks. Customers also use wide area networks to interact with applications that comprise the products and services that they are buying. (Wide area networks are principally telecommunications-based, but can include postal and physical networks.)

Table 13.3
Processes/Systems Related to the Customer Interface

Function, Element, or Entity	*Processes*
Business direction and support	Creation and monitoring of strategy Organization support and development Information systems management
Market/customer relations	Includes marketing and intermarketing Creation and management of products Part of obtaining orders, implementing orders, and providing continuing support Marketing, sales and relationship management, project implementation, especially through the following:
Call center or similar	Part of obtaining orders, implementing orders, and providing continuing support (Outbound) proactive marketing

Function, Element, or Entity	Processes
Making the product work for customers	Part of obtaining orders, implementing orders, and providing continuing support Processes of installation, billing, fault rectification, upgrades, user help relating to the following:
Actual products and services	Customer-facing information applications... used through a wide area network interface
Integrated platforms	With common look and feel to the customers (and employees) and with effective transaction processing through client-server/hub-and-spoke/online transaction processing, switches, store-and-forward devices, etc.
Technology and information elements	Network, computing, terminal equipment, software elements

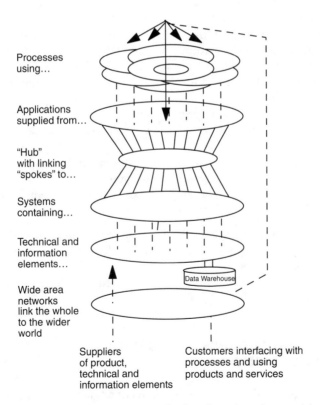

Processes using...

Applications supplied from...

"Hub" with linking "spokes" to...

Systems containing...

Technical and information elements...

Data Warehouse

Wide area networks link the whole to the wider world

Suppliers of product, technical and information elements

Customers interfacing with processes and using products and services

Figure 13.2 IS architecture for telecommunications-based service provision.

Technical and information elements from suppliers also come to the telecommunications-based service provider through wide area networks.

Figure 13.2 shows the layers of application, hub-and-spoke, and systems that join the two. In accordance with Section 13.3, the diagram also serves directly to show how the telecommunications-based service provider's "internal" people, processes, and applications link to the underlying "internal" systems and technical and information elements.

It is unlikely, however, that the whole business will stand still whilst the grand information systems plan is created, the applications is developed in parallel with perfect interfaces and no data conflict, then implemented simultaneously with instantaneous, perfect results. Some realistic and practical barriers might include the following:

- Process areas will certainly not know what they actually want until they get started. It is highly unlikely that any significant information system has ever gone into use with no changes in the original user specification.
- Process areas will develop at varying speeds. Thus the input and output models will not be ready on time.
- Proprietary packages will be used to support many of the processes and activities (e.g., accounting, contact management). Even if all operate, say, under a recognized standard operating system, they'll probably have individual characteristics that make them incompatible.
- There will be existing legacy systems.
- Customers and suppliers will have their own approaches that may not match.
- There may be a mixture of online systems and batch-processing systems (this happened in a recent bank merger).
- There is unlikely to be overall consensus on requirements and priorities.

A practical approach is to define overall information systems objectives, standards, and procedures, but to break down the overall application into manageable portions.

The next section illustrates how this might be done, process area by process area.

13.6 MAPPING PROCESS AREAS WITH APPLICATIONS/SYSTEMS

The information applications and systems of a telecommunications-based service provider are used in one, some, or all of the process areas.

Table 13.4 shows which process area should be given *lead user* responsibility.

Each application/system is also mapped in the process outline and the actual use of each is described in the text of each of the seven preceding chapters. "Lead user responsibility" and other aspects relating to the development and management of these applications, and to a consistent overall architecture and standard, are covered immediately afterwards:

13.7 INFORMATION SYSTEMS MANAGEMENT MODEL FOR A TBS-P

The effective development and subsequent operation of information systems for a telecommunications-based service provider requires the alignment of three distinct aspects: business objectives, user functions, and technical development and operation. To achieve this alignment and achieve a sound practical result, the tripartite organization shown in Table 13.5 is recommended. The roles and responsibilities of the IS project/operations department are based upon the role of the systems integrator described in Chapter 9.

13.8 THE WORK OF THE IS PROJECT/OPERATIONS DEPARTMENT

The tasks listed in column 4 of Table 13.4 above are the standard ones for most information systems management teams. Further detailing is really beyond the scope of this book. However, here are some examples and pointers that are particularly relevant.

13.8.1 Technology

There are continuous advances in technologies supporting the customer interface. Switchboards give way to *automatic call distribution* (ACD), intelligent/skill-based routing approaches, and customer self-service (particularly using IVR). Voice recognition is making rapid progress. Reduced costs of telephony and competitive pressures are making call centers both bigger and more geographically remote from the customers they serve.

The steady increase in importance of the *call-center* approach to business organization is building upon the (itself relatively recent) *client-server* approach, so that the focus of computing systems is on hub-and-spoke technology mixed with computer telephony integration. On-line transaction processing is almost essential for telephone banking and similar applications.

Table 13.4
Lead Process Areas Versus Applications/Systems

Application/system	Chapter 6 Strat	Chapter 7 Markt	Chapter 8 Prodt	Chapter 9 Ordr	Chapter 10 Implt	Chapter 11 Serve	Chapter 12 Suppt	Chapter 13 IS
Strategic analysis resources	Lead							
Management accounting systems	Lead							
Organization reporting systems								
Marketing communication including intermarketing, Web site authoring, etc.		Lead						
Sales lead generation		Lead						
Marketing collateral—brochures, etc.		Lead						
Market research resources		Lead						
Product: All internal and external information and physical elements bought in and/or supplied to customers			Lead					
Product information—generic terms, conditions and pricing, configuration, etc.			Lead					
Product support systems—order input, billing, etc.			Lead					

Notes: Strat: Strategy and Monitoring; Markt: Marketing and Intermarketing; Prodt: Product Creation and Management; Ordr: Obtaining orders; Implt: Implementing Orders; Serve: Continuing Service; Suppt: Organization Support and Development; IS: Information Systems.

Application/system	Chapter 6 Strat	Chapter 7 Markt	Chapter 8 Prodt	Chapter 9 Ordr	Chapter 10 Implt	Chapter 11 Serve	Chapter 12 Suppt	Chapter 13 IS
Outbound telephony: telemarketing—scripts, predictive dialers, etc.				Lead				
Selling systems—"prospect to order" tracking, bid preparation, sales management				Lead				
Implementation project management					Lead			
Logistics systems—stores, delivery					Lead			
Inbound telephony: PABX lead user customer service system—inbound call handling						Lead		
Customer and employee training systems						Lead		
Financial accounting systems including automated payment systems, banking, etc.							Lead	
Individual customer contracts							Lead	
Personnel systems, payroll							Lead	
Quality management system, procedures							Lead	
Building, vehicle records, fixed asset register							Lead	

Table 13.4 (Continued)

Application/system	Chapter 6 Strat	Chapter 7 Markt	Chapter 8 Prodt	Chapter 9 Ordr	Chapter 10 Implt	Chapter 11 Serve	Chapter 12 Suppt	Chapter 13 IS
Audit and security, disaster recovery, insurance							Lead	
Information systems administration, assignment of equipment, licenses, etc.								Lead
PABX call plan, IS supplier information								Lead

Table 13.5
IS Management Systems—Roles and Responsibilities

Team	IS Steering Team	User Group	IS Project/ Operations Dept.
Role	Define business objectives Oversee standards, architecture etc. Provide resources Set priorities Agree project plans Review projects Agree supplier selections Agree common "look and feel" (marketing/ intermarketing)	Functionality Usability Performance required to meet the assigned tasks Prototypes Coordinates user concerns, ideas, and other inputs Agrees on project plans Agrees on training plans and oversees its effectiveness	System design Implementation design Integration Manages procurement process Contract management supplier relationships Integration Installation, commissioning Training System operation Program office
Scope	Business as a whole	Single application or group of applications	Single system or group of systems
Members	CEO (chair) VP IS (secretary) Process area heads Other senior managers as required User representatives by invitation	Senior representative from process area having "lead user responsibility" Other representatives of users, from a range of role, rank, and function	Project manager
Meets	Monthly or as required for major decisions or milestones	Intense activity at start of project, then as required by project plan	Operates continuously

One result of CTI is that screens of customer data follow the customer call as it is handled by one agent and then another. A second result is that customer data is readily available to each agent during the call.

Integrated messaging brings together text, fax, voice, data—giving whatever presentation is most convenient.

The Internet is changing at a remarkable rate. Its user base is growing at 10% per month. Its technologies are becoming ever more into the mainstream, with the increasing occurrence of private Internets for individual corporations (intranets)—with internal World Wide Webs, message post offices/exchangers, search engines, and other features. Intranets will almost invariably be connected to the Internet proper.

Structured cabling plans and system built approaches to flooring, partitioning, and power supplies are essential in deploying systems for the fast-changing environment of the telecommunications-based service provider. Professional advisers and experienced contractors must be selected to deliver the most suitable solution.

13.8.2 Security

An effective and well-enforced system of information backup is essential to the very survival of a telecommunications-based service provider.

The increased penetration and connectivity of corporate systems mean that computer security is a key issue. Information can be stolen more easily. Similarly, it has never been easier for an individual person to create and release documents to the wider world. Firewalls are needed—servers that prevent unauthorized users or messages coming on to the corporate network. Procedures and rules have to be created and followed—for example, to forbid the use of floppy disks without them going through a virus checker first. There has been a rush of literature and advice on this subject.

13.8.3 Skills and Supply Issues

Given the explosion of subtechnologies and specialist areas that is the new telecommunications-based age, there is a need for both a focus on business needs and a good level of skill and knowledge in the use of specialists. This twin requirement can be met by the use of outsourcing for parts and/or a systems integrator for an overall solution. In any case, it is advisable to "market test" what, at first sight, may to be something better to do in-house. What is quoted by a specialist outsider will probably give a clear picture of the real costs of doing something in-house.

In the telecommunications-age there is one organization that exemplifies this approach above all others—the telco. Over the past few years, the emphasis in this area of technology has oscillated between customer premises equipment and network-based functionality. Where a function is in the network of a telco (or embedded in the Internet in the case of the "network computer"), then what the end user will experience is a service.

13.9 INTERNET TECHNOLOGIES

Increasingly, firms are basing their internal information systems strategies upon Internet technologies. There are several reasons for this:

- The current focus is there, so that is where knowledge and interest are greatest.
- Internet technologies are built around telecommunications and sharing information widely amongst enterprises and markets, rather than older single enterprise systems.
- Internet approaches fit neatly with hub-and-spoke approaches.
- Internet approaches are multimedia.
- Internet economics reflect the steady downward costs of telecommunications (e.g., the rationale of the network computer).
- Internet enables a single point for application software—easier to manage and update.
- Internet/hypertext approaches now developing on the WWW are more intuitive and much more fun than "double backslash Ctrl XYZ space space colon" commands.

However, there are obvious dangers in opening up the information resources of the corporation, as outlined in Section 13.8.2. Internet was started to be anarchic and open; many of its underlying philosophies remain so.

13.10 CONCLUSIONS

As for most other IS teams, the telecommunications-based service provider must ensure the following:

- A clear vision of the business that is reflected in the attention and resourcing devoted to IS;
- A well thought through overall IS architecture;
- Disciplined working of the management arrangements described in Table 13.4 above;
- Clear standards, properly enforced;

- Well-documented processes;
- Curiosity and the ability to absorb all that is changing—weekly—in the technological environment, but discipline to embrace only that which is relevant to the business proposition;
- IS team with hands on experience of the core technologies used;
- Where that experience has not been obtained, it is likely that specialist temporary assistance can be found. Outsourcing is a worthwhile option to consider for most areas of IS.

And it is this last insight that brings us full circle in this chapter. The tasks that one organization is "outsourcing" are the same ones that are being sold back by another organization—in the role of *service provider*.

This completes third portion of this book—the practice of telecommunications-based service provision. Chapter 14 contains a summary and conclusions.

Conclusions and Actions   **14**

14.1 RAISON D'ÊTRE OF SERVICE PROVISION

There can be little doubt that the information age in which we live has already had a fundamental effect on the way that we live and the way we do business. The telecommunications available today are, in the author's opinion, the principal driver of the mid1990s acceleration of the information revolution, though the revolution probably started 10 or even 15 years ago. There is an enormous opportunity for the businesses that know how to use telephony—or know a service provider who can help them.

The technologies of the telecommunications age have altered the effects of distance and of time. Instant communication has proven to be a melting pot of national differences—for example, reckoned by many to have hastened the reuniting of East and West Europe. Transaction-intensive enterprises such as banks, insurance companies, phone companies, and travel companies have all merged in one way or another somewhere around the globe. Increased use of telephony as a channel from customer to supplier has thrown up telecommunications-based service provision as a critical issue for airlines and banks, to note two particular examples.

The technology picture of this age is both divergent and convergent. Although IT has exploded into many subareas, each with its requirements for specialization, it has also become the arena of any possibility—any client/any network/any bit.

Technology has clearly advanced faster than the ability to organize it usefully. There is massive excess of choice of course of action and, especially, information. Reportedly, there is more information in a single copy of the *New York Herald* than was encountered in the whole lifetime of a typical 17th century person. There is a need to be selective. There is a need for gatekeepers who can direct the relevant items of what is available to what meets our individual needs and interests. Technology has become unthinkably complex. A very high proportion of people cannot use more than a fraction of the functionality of a PC or of a video cassette recorder (VCR).

There is an explosion of choice of supplier. Mobile phone technology may be easier to use, but there is a bewildering choice of terminals and services for the average person, who may also feel that the overall sales proposition is sometimes engineered against him—for example, with excessively complex and restrictive contracts. The need is for a service provider who can be trusted, whose brand is above suspicion, who will seek to make an honest profit from a relationship rather than a single transaction.

14.2 CONCEPT OF SERVICE PROVISION

The telecommunications-based service provider is essentially a channel between source content and technical devices on the one hand and customers on the other. The value proposition for both parties is that they will benefit from a service that brings them together in a convenient form, without the necessity of carrying out tasks beyond one's immediate desire, knowledge, or capability. The function of the service provider is to act as an intermediate value chain, as shown in Figure 14.1.

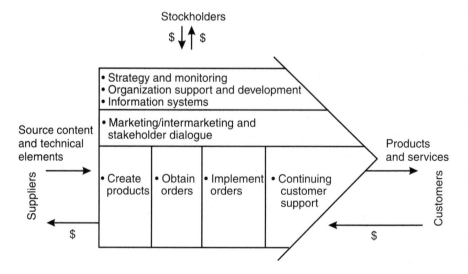

Figure 14.1 Service provider value chain.

Although telecommunications-based service provision has been described primarily as making products and services from information and technology elements taken from a third party, it is equally valid to consider it as constituting the customer interface for products or services from its own organization (e.g., the reservations line of an airline). There can be few transaction-intensive, information-based organizations that will not benefit from such an arrangement.

The remarkable advances of the Internet, powered by the WWW, browsers, Java applications, and increased robustness of operation, have been matched by raw enthusiasm and entrepreneurial flair. The Internet proper has suffered at times from unacceptable slowdowns, but these are being overcome by the introduction of several Internets and intranets—each optimized to specific tasks or costing models.

14.3 PRACTICE OF SERVICE PROVISION

The central purpose of this book has been to propose how to operate a telecommunications-based service provider. The approach has been to show the operation as a whole and to explain the functioning of each part. The concept of the eight process areas has been used throughout the book. The aim has been to use common vocabulary and formats. Furthermore, the process areas have been described in order of the chronology of a strategy/product/sale/implementation.

The resulting list of processes is shown in Table 14.1.

Table 14.1
Telecommunications-Based Service Provision—Processes

Process Area	*Processes*
Strategy and monitoring	Analyze strategic situation
	Identify and select from choices
	Set overall objectives
	Create and publish detailed plans
	Monitor results and feed back into strategy
Marketing/intermarketing	Positioning the product and the brand
	Generating sales leads and supporting sales channels
	Measuring perceptions; managing stakeholder strategic dialogue
	Central information function (part of)

Table 14.1 (Continued)

Process Area	Processes
Product	Concept (prototype may occur here) Business case and technical feasibility Design Prototype, trials, and development Launch Sale, delivery, and operation In-service development Withdrawal
Obtain order	Classic sales process comprises: Preparation Inquiry Presentation Close Administration Processes and channels vary by customer type: large, specialist/niche, volume market
Implement order	Kickoff meeting Procure required information or technology elements to build service Assign people to build and deliver service, training, etc. Advise customer of progress and expected delivery—offer alternatives or enhancements if required or appropriate Assemble and integrate elements of the product Deliver to the customer Train the customer Obtain customer acceptance certificate
Continuing service	Accept handover of the customer from the "implement orders" team Initiate the "customer service relationship" Provide "help desk support" Bill the customer Manage technical faults and functional problems Carry out shifts and changes Provide enhancements Measure/conduct dialogue/analyze... and act upon what is found Discontinue service

Process Area	Processes
Organization support and development	Financial accounting, cash management, taxation Company secretarial Commercial, regulatory (handled in "strategy") Procurement and logistics Human resources management (HRM), recruitment, payroll, etc. Quality management systems Team secretarial and administration Facility management Business continuity: audit and security, disaster recovery, insurance
Information systems	Set business objectives for IS Define scope of IS and architecture to match Assign applications/systems to relevant process areas Develop and manage systems

A cultural approach has been suggested that reflects the increased need of employees to do meaningful work and to provide good services to customers. No modern organization functions without information systems. An architecture for telecommunications-based service provision systems has been articulated, described, and illustrated in the previous chapter. Figure 13.2 linked together the eight process areas and showed how they mapped onto product and internal information systems and information elements. It is repeated here as Figure 14.2.

14.4 WHAT ACTIONS NOW?

Some people who have read this book will go on to create all or part of a telecommunications-based service provider. Some will already work in what they will easily recognize to be a telecommunications-based service provider. The use of the eight-process-area approach will provide an opportunity to ensure that every necessary process or task is accounted for at the start and is mapped correctly. Experience suggests that all of the tasks listed will need to be carried out, even if not intensively at first.

Some readers of this book will be specialists in one or more of the process areas described. It is to be hoped that they will have benefited from broadening their understanding of the other areas. It is to be hoped

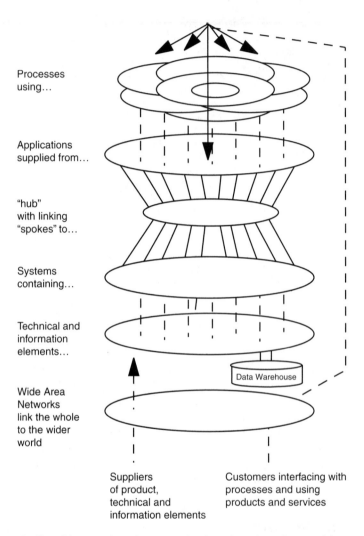

Processes
using...

Applications
supplied from...

"hub"
with linking
"spokes" to...

Systems
containing...

Technical and
information
elements...

Data Warehouse

Wide Area
Networks
link the whole
to the wider
world

Suppliers
of product,
technical and
information elements

Customers interfacing with
processes and using
products and services

Figure 14.2 IS architecture for telecommunications-based service provision.

that even specialists will have found value from the articulation of what their area is and does. Hopefully, too, they will have gained one or two new items of knowledge or just ideas.

The development and operation of information systems will proceed better from a common vision of the organization's processes and a picture of how the systems will interact. It is unlikely that each system will be directly connected to the next at the start, but, at the very least, each system should operate to strict rules of format for, say, customer names. The proposed tri-partite method of managing systems should save time otherwise spent regretting the absence of the user spec after the system has already been built.

14.5 NEW HORIZONS

Finally—partly for planning, but partly for fun—here are some predictions for the further development of telecommunications-based socio-technologies and the Internet in particular:

- Costs of bit transport, processing, and storage will continue to fall steeply.
- Power of computing will continue to rise steeply.
- Devices and services will continue to become easier to use.
- Voice has been a dominant method of real-time interaction throughout history. Improving voice recognition technology will keep voice at the fore.
- The Web's intuitive access to information will progressively become the data interface standard, as it already is on the Internet. Many Web sites will be created that are gateways to conventional databases (e.g., WWW.bookshop.co).
- The *Internet* will become an increasingly *generic* term to the naive user; Internet technology will be used within an increasing proportion of overall solutions.
- Service provision, as a channel, has given governments an opportunity to force competition into previously monopolistic areas. This tendency will continue—especially in newly privatizing environments. Dominant network operators will be required to show no undue preference to their own in-house service provision functions.
- There will be low-price network computers widely available. This means that a very large number indeed of applications will be available at will.
- Cable modems (27 million bits per second) will become widely available.
- Sleaze will be policed off the Internet.
- Commercial transactions on the Web will be the norm in United States first. Other places will struggle to keep up.
- There will be one or more security crises that will rival the Wall Street crash, before the situation stabilizes.
- Electronic messaging, with gateways and easy to use text-voice-text conversion will overtake physical postal message services. Hard

copy will be delivered to the most convenient place, especially through messaging exchanges.

- Phone companies will bring more and more intelligence away from customer sites and into their networks. The implication is that *services* will be made much more attractive than *equipment* for most applications.
- The normal way to set up in the information business will be to write an application in Java, put it on a Web site, get people using it, and sell them enhancements and upgrades.
- GSM will become the world roaming standard for mobile phones, with multistandard interfaces common to DECT and satellite.

Some of these are speculative. Whatever happens, the explosion of technologies and information will continue. The need to select, channel, and deliver these to customers will continue and increase. Although user interfaces will become easier, particularly where voice is the medium, customers will be prepared to spend ever less time in learning about them.

Expectations of customers for customer service will continue to rise—so could costs. But the price of services will continue to fall relative to overall wealth, so demand will increase. The need for helpful, knowledgeable service providers with access to the best source content and technical elements is set to increase more and more. All information-based businesses need to organize behind a telephony front end and integrate systems, processes, and behaviors accordingly.

Bibliography

Bahrami, H., "The Emerging Flexible Organization: Perspectives From Silicon Valley," *California Management Review,* Summer 1992, Vol. 34, No. 4.

Bahrami, H., and S. Evans, "Flexible Recycling and High-Technology Entrepreneurship," *California Management Review,* Spring 1995, Vol. 37, No. 3.

Bartholomew, M. F., and J. Harris (Eds.), *BT UK Performance Report*, BT. London, 1989

Bartholomew, M. F., "Theory T" - a simple, solid framework for a teleworking environment," *Teleworking Magazine*, London, Aug. 1992.

Bartholomew, M. F., "Keeping Up the Pace," *TQM Magazine,* Bradford, UK: MCB University Press, Dec. 1993.

Bartholomew, M. F., "Challenges Facing Service Providers for the 1990s," *AIC Conferences*, London, 1994.

Canter, L. A., and M. S. Siegel, *How to Make a Fortune on the Information Superhighway*, New York, NY: HarperPerennial, HarperCollins, 1995.

Coulson-Thomas, C., *Transforming the Company*, London: Kogan Page, 1992.

Dicken, P., *Global Shift,* (3rd ed.), London: Paul Chapman Publishing, 1995.

Dickinson, S., *How to take on the Media,* London: Weidenfeld & Nicolson, 1990.

Ellsworth, J. H., and V. Matthew, *Marketing on the Internet*, New York: John Wiley and Sons, 1995.

Ericsson, D. (Ed.), *Virtual Integration*, Stockholm, Sweden: Unisource, 1996.

Fairclough, K., and R. Taylor, *Knowledge Management - A Framework*, London: Unisys Corporation, 1995.

Gates, B., with N. Myhrvold and P. Rinearson, *The Road Ahead*, New York, NY: Viking Penguin, 1995.

Hallows, R., *Service Management in Computing and Telecommunications*, Norwood, MA: Artech House, 1995.

Hastings, C, P. Bixby, and R. Choudry-Lawton, *The Superteam Solution*, Aldershot, U.K.: Gower, 1986.

Heath, W., *Wired Whitehall 1999*, London: Kable, 1995.

Johnson, J., and K. Scholes, *Exploring Corporate Strategy*, (3rd ed.), Hemel Hempstead, U.K.: Prentice Hall, 1993.

Minoli, D., *Telecommunications Technology Handbook*, Norwood, MA: Artech House, 1991.

Obeng, E., and S. Crainer, *Making Re-Engineering Happen*, London: Ashridge/FT Pitman Publishing, 1994.

Paetsch, M., *Mobile Communications in the U.S. and Europe*, Norwood, MA: Artech House, 1993.

Patel, K. J, *Who Knows Wins*, Bristol, U.K.: Jordans, 1996.

Savage, C. M., *5th Generation Management*, Digital Equipment Corporation, USA, 1990.

About the Author

Martin Bartholomew, the managing director of Periphonics (Voice Processing Systems) Ltd., with responsibility for Europe, Africa, and the Middle East, is based in the United Kingdom but travels extensively around the world. Periphonics is a specialist supplier of interactive voice response technology and has worldwide annual revenues of around $100 million. The corporate headquarters is in Long Island, New York.

Martin has 22 years experience in telecommunications, cable, and information systems. He has worked in three European telcos in general management and business development posts. He worked in BT Account Management and Marketing Communications for five years before moving to Mercury as director, Government Services and then director, Mobile Services. He was CEO of Latvia's PTO (a joint venture of the Latvian Government, Cable & Wireless and Tele-Finland) immediately after its transformation into a joint stock company in 1994.

From 1966 to 1974, he qualified and served as an aircrew officer in the Royal Navy. He subsequently gained an M.A. in management from the University of Kent. He is a Fellow of the Institute of Management, a member of the Institute of Directors and of the Strategic Planning Society, and a Visiting Fellow on the University of Luton MBA program.

Martin has spoken frequently on strategy, quality, voice processing, and customer service and has had a number of papers and articles published in these areas.

Index

The Artech House Telecommunications Library

Vinton G. Cerf, Series Editor

For further information on these and other Artech House titles, contact:

Artech House	Artech House
685 Canton Street	Portland House, Stag Place
Norwood, MA 02062	London SW1E 5XA England
617-769-9750	+44 (0) 171-973-8077
Fax: 617-769-6334	Fax: +44 (0) 171-630-0166
Telex: 951-659	Telex: 951-659
email: artech@artech-house.com	email: artech-uk@artech-house.com
WWW: http://www.artech-house.com